A Text Book Of

EMERGING TRENDS IN CIVIL ENGINEERING

Course Code 22603

Semester - VI

THIRD YEAR DIPLOMA IN CIVIL ENGINEERING GROUP

As Per MSBTE's 'I' Scheme Syllabus

MUKESH NARENDRA GANGRADE

M.E. (Water Resource Engineering), B. E. (Civil)
H.O.D. Civil Engineering Department
Vidyavardhini's Padmshri Bhausaheb Vartak Polytechnic,
Vasai Road (W), Dist. Palghar

PRATHMESH VILAS PATIL

M.E. (Construction Engineering & Management), B. E. (civil)
H.O.D. Civil Engineering Department
Alamuri Ratnamala Institute of Engineering and Technology
Shahapur, Dist. Thane

NIRALI PRAKASHAN
ADVANCEMENT OF KNOWLEDGE

N4557

Emerging Trend in Civil Engineering **ISBN 978-93-89825-62-6**

First Edition : January 2020

Published By :
NIRALI PRAKASHAN
Abhyudaya Pragati, 1312, Shivaji Nagar
Off J.M. Road, PUNE – 411005
Tel - (020) 25512336/37/39, Fax - (020) 25511379
Email : niralipune@pragationline.com

➢ DISTRIBUTION CENTRES

PUNE

Nirali Prakashan (For orders within Pune) : 119, Budhwar Peth, Jogeshwari Mandir Lane, Pune 411002, Maharashtra
Tel : (020) 2445 2044, Mobile : 9657703145
Email : niralilocal@pragationline.com

Nirali Prakashan (For orders outside Pune) : S. No. 28/27, Dhayari, Near Asian College Pune 411041
Tel : (020) 24690204 Fax : (020) 24690316; Mobile : 9657703143
Email : bookorder@pragationline.com

MUMBAI

Nirali Prakashan : 385, S.V.P. Road, Rasdhara Co-op. Hsg. Society Ltd.,
Girgaum, Mumbai 400004, Maharashtra; Mobile : 9320129587
Tel : (022) 2385 6339 / 2386 9976, Fax : (022) 2386 9976
Email : niralimumbai@pragationline.com

➢ DISTRIBUTION BRANCHES

JALGAON

Nirali Prakashan : 34, V. V. Golani Market, Navi Peth, Jalgaon 425001, Maharashtra,
Tel : (0257) 222 0395, Mob : 94234 91860; Email : niralijalgaon@pragationline.com

KOLHAPUR

Nirali Prakashan : New Mahadvar Road, Kedar Plaza, 1st Floor Opp. IDBI Bank, Kolhapur 416 012
Maharashtra. Mob : 9850046155; Email : niralikolhapur@pragationline.com

NAGPUR

Nirali Prakashan : Above Maratha Mandir, Shop No. 3, First Floor,
Rani Jhanshi Square, Sitabuldi, Nagpur 440012, Maharashtra
Tel : (0712) 254 7129; Email : niralinagpur@pragationline.com

DELHI

Nirali Prakashan : 4593/15, Basement, Agarwal Lane, Ansari Road, Daryaganj
Near Times of India Building, New Delhi 110002 Mob : 08505972553
Email : niralidelhi@pragationline.com

BENGALURU

Nirali Prakashan : Maitri Ground Floor, Jaya Apartments, No. 99, 6th Cross, 6th Main,
Malleswaram, Bengaluru 560003, Karnataka; Mob : 9449043034
Email: niralibangalore@pragationline.com

Other Branches : Hyderabad, Chennai

niralipune@pragationline.com | www.pragationline.com

Also find us on www.facebook.com/niralibooks

*

This book is Dedicated to my mother ...

Smt. Sunanda Narendra Gangrade

Author

Mukesh Narendra Gangrade

mngangrade@gmail.com

*

*

This book is Dedicated to my mother and father ...

Mrs. Kamini Vilas Patil & Mr. Vilas B. Patil

Author

Prathmesh Vilas Patil

prathmeshpatil19@gmail.com

*

Preface ...

We are presenting this book entitled as A Text Book of **"Emerging trends in Civil Engineering"** for the students of polytechnic colleges preparing for **Diploma in Civil Engineering Semester VI** as per **'I' scheme of M.S.B.T.E.**. The object of this book is to present the subject matter in the most concise, compact, to the point and in lucid manner.

"Emerging trends in Civil Engineering" subject is offered by MSBTE as Subject to the Third Year Civil Engineering Group. Civil Engineering sector has completed number of projects with conventional techniques to meet the needs of the society. But, in recent two decades, various new innovative techniques are being used worldwide, which our practicing engineers, are also partially using to achieve their goals. The emerging trends in Civil Engineering help to complete the undertaken projects within prescribed schedule, saves the natural resources and to make the projects eco-friendly. This subject helps to make awareness about soft computing techniques, new materials, advanced machineries, sustainable resource management and advancement in Civil Engineering. A separate appendix is prepared to cover list of microproejcts.

We are very much thankful to the publisher **Shri. Dineshbhai Furia** and **Shri. Jignesh Furia.** We also thank to Mr. Santosh Bare, Mr. Shashikant Patel, Mrs. Anagha Kaware (Co-ordination and Proof Reading), Mrs. Yojana Deshpande (Figure Drawing) and staff of Nirali Prakashan for bringing out this book.

Any undetected and unintentional error, omissions, suggestions etc. from students and teachers for improvement brought to our notice in good spirit are most welcome.

M. N. Gangrade

P. V. Patil

Acknowledgement

I am thankful to my wife, **Mrs. Rajani M. Gangrade** for their support while writing this book.

Author

Mukesh Narendra Gangrade

mngangrade@gmail.com

I am thankful to my lovable parents, **Mr. Vilas B. Patil** and **Mrs. Kamini V. Patil** for their blessing. Also thankful to my brother **Mr. Ashish Vilas Patil**, my sister and brother-in-law **Mr. & Mrs. (Dr.) Puja Hitendra Raul** for giving me moral support to write and present this book.

I am also thankful to **Prof. M. N. Gangrade** (H.O.D. Civil Engineering Department, Vartak Polytechnic) and **Dr. L. S. Bothra** (Chairman of **A.R.M.I.E.T.** college, Shahapur) for their support and moral inspiration to write and to present this book.

Author

Prathmesh Vilas Patil

prathmeshpatil19@gmail.com

Syllabus ...

1. Soft Computing Techniques (Marks 12, Hours 08)

1.1 Introduction of soft computing techniques and its types,

- Merits and demerits of soft computing technique
- Graphical User Interface Software (GUI) .

1.2 Introduction, salient features and applications of software's

- REVIT, ETAB, 3D Architect Home, Build-Master, HEC-RAS, STRAP, WaterGEMS, Tekla, ArcGIS, QuikGrid, STAAD.Pro, SAFE, RISA-Connection, Civil 3D, Site 3D, SkyCiv Structural 3D, SAP 2000, MIDAS, LUSAS, BricsCAD, Estimate Master, ProEst, WinEst, Clear Estimate, Procore, Buildertrend, Building Management System (BMS), Plant Design and Management System (PDMS), Building Information Modeling (BIM), Primavera Pro, Microsoft Project (MSP)

2. Recent Construction Materials (Marks 14, Hours 10)

2.1 Building Materials

- Artificial sand, Sensi tile, carbon fibre, Bricks made up of cigarette butts, 3D printed bricks, Translucent wood, laminated timber, 3D Tiles.

2.2 Road Materials

- Geo-synthetics, Noise-reducing asphalt, Porous Pavement, Plastic Roads, solar roads, Anti Icing Roads, Piezoelectric roads.

2.3 Concrete Materials

- Portland Pozzolana Cement, Portland Slag Cement.
- New admixtures - Masterglanium, Polycarboxylic Ether.
- Self Healing Concrete, Fibre-Reinforced Concrete, High Strength concrete, High Performance Concrete, Nano concrete, Light transmitting concrete.

2.4 Sustainable Materials

- Ground Granulated Glass Blast-furnace Slag (GGBS) concrete, Aero-gel insulation, Cooling bricks, Green concrete, Timbercrete, Ferrock.

3. Latest Tools and Equipments (Marks 14, Hours 10)

3.1 Survey equipment

- LiDAR, Scan Station, Global Positioning System (GPS), Geographical Information System (GIS), Photogrammetry, Drones, Direct Reading Grade Rods, 3D Laser scanning, laser level.

3.2 Construction Equipment

- Earth moving equipment- Skid and crawler loaders, trenchers, scrappers, wheeled loading shovels, advanced plastering machine, Bridge launcher.

3.3 Material handling equipment

- Cranes, conveyors, hoists, forklifts, mobile concrete mixer, paver, road header, tunnel boring machine.

4. Sustainable Resource Management (Marks 12, Hours 08)

4.1 Water resource management

- Principles and techniques

4.2 4R's in waste management

- Reduce, Reuse, Recycle and Recover
- Concept of Zero Waste

4.3 Reuse of waste in construction

- Fly Ash, Slag and Plastic

4.4 Renewable energy sources

- Solar energy, biomass, wind energy, ocean wave energy, Geo thermal energy and hybrid power system.

4.5 Energy Audit

- Necessity and methods.

4.6 Natural disaster management

- Flood, Earthquake, Tsunami, Volcanic Eruption, Hurricanes, Landslides.

4.7 Site Safety

- Necessity, principles, tools, techniques, laws, rules and regulations.

5. Advancement in Construction (Marks 18, Hours 12)

5.1 Building construction techniques

- Zero energy building, Green building, Mass housing-precast housing, prefab homes, pre-engineering building, Solar Paints, Building Integrated Photovoltaic (BIPV), Earthquake Resisting Controls-Isolation and Dissipation.

5.2 Road construction techniques

- 3D Printing, Road Printer, smart roads

5.3 Coastal construction techniques

- Sound Proofing walls, water resistant roofs, high performance doors and windows, air and moisture barriers.

5.4 Ground improvement techniques

- Advanced piling techniques - Stone Column, Vibro Floatation, Micro Piles, Soil Nailing
- Vertical drains-Sand Drains, Pre-Fabricated Vertical Drains,
- Thermal Methods- soil heating and soil freezing.

❑❑❑

Contents ...

❑❑❑

Chapter **1**

Soft Computing Techniques

Weightage of Marks = 12, Teaching Hours = 08

Syllabus

1.1 Introduction of soft computing techniques and its types,
   - Merits and demerits of soft computing technique.
   - Graphical User Interface Software (GUI).

1.2 Introduction, salient features and applications of software's - REVIT, ETAB, 3D Architect Home, Build-Master, HEC-RAS, STRAP, WaterGEMS, Tekla, ArcGIS, QuikGrid, STAAD. Pro, SAFE, RISA-Connection, Civil 3D, Site 3D, SkyCiv Structural 3D, SAP 2000, MIDAS, LUSAS, BricsCAD, Estimate Master, ProEst, WinEst, Clear Estimate, Procore, Builder trend, Building Management System (BMS), Plant Design and Management System (PDMS), Building Information Modelling (BIM), Primavera Pro, Microsoft Project (MSP).

About this Chapter

At the end of this chapter, students will be able to:

- Enlist the characteristic uses of soft computing techniques in civil engineering.
- Identify merits and demerits of soft computing techniques.
- Understand the general features of software's used in civil engineering.
- Suggest the suitable software/s for the given construction activity.

1.1 INTRODUCTION OF SOFT COMPUTING TECHNIQUES AND ITS TYPES

- Construction industry is highly competitive and faces challenges in the areas of costs of projects, delays in construction activities, labour productivity, disputes, tenders, bidding prices, safety aspects, rate of materials, maintenance costs, risk analysis etc. which are highly complicated in nature.
- To deal with these challenges, Artificial Intelligence (AI) techniques like fuzzy logic, case-based reasoning, probabilistic methods for uncertain reasoning, classifiers and learning methods, Artificial Neural Networks (ANN), Genetic Algorithms and hybrid techniques are widely used in the field of Construction Management (CM).
- In the last two decades of twentieth century, there was a surge in publications dealing with Artificial Intelligent techniques and especially ANN in various aspects of CM. In 2001, Adeli and Yeh provided a comprehensive review of such applications made before the turn of the century.
- The current work presents a review of about 70 papers published in the area of CM. The objective of the paper is to highlight the applications of ANN in the following fields of CM: Cost, Productivity, Risk Analysis, Safety, Duration, Dispute, Unit rate and Hybrid Models. Further critical review of the findings will help the readers to focus on important areas for potential use and development of ANN in the said areas of CM.
- The future scope will facilitate continued research efforts. The paper is further synthesized as follows: Initially a brief introduction on ANN is presented and is followed by the assessment of their recent applications in the areas of Cost, Productivity, Risk Analysis, Safety, Duration, Dispute, Unit rate and Hybrid Models.

- Discussion and critical review is done in the preceding section followed by author's comments on the findings and future scope.
- Civil Engineering is a diversified field ranging from water resources to design and analysis of structures.
- The nature of the problems encountered in this field is of a complex nature and mostly requiring human intervention in the form of past experience and intuition.
- The heuristic nature of problems poses challenges to a civil engineer thereby making him resort to time and resources saving computational tools.
- Soft computing techniques viz., Neural Networks, Fuzzy Logic and Genetic Algorithms either working independently or complementing each other support the engineering activities by harnessing the cognitive behaviour of the human mind to arrive at cost effective solutions.
- The review paper is an attempt to highlight the applications of Artificial Neural Networks and Genetic Algorithms in the field of Civil Engineering.

1.1.1 Merits and demerits of soft computing technique

Merits of soft computing technique

- **Saves time**: When you are using the soft computing techniques, it will save your time and you can make better and more efficient designs in shorter time duration.
- **Easy to edit**: When you are making designs, you may find the need to make alterations. When you are using soft computing techniques, it will be much easier to make any changes because you can fix the errors and modify the drawings easily.
- **Decrease in error percentage**: As the soft computing techniques makes use of some of the best tools, the percentage of error that occurred because of manual designing is significantly reduced.
- **Decrease design effort**: When it comes to the amount of effort that was needed for the sake of designing the different models, it has been reduced significantly because the soft computing techniques automates most of the task.
- **Code re-use**: As the entire task is carried out with the help of computer tools, it removes the problem of duplication of labor, you can copy the different parts of code and design which can then be reused multiple times over and over again.
- **Easy to share**: The soft computing techniques tools make it easier to save the files and store it in a way that you can use it time and again and send it without any unwanted hassles too.
- **Improved accuracy**: There is absolutely no doubt about the fact that the kind of accuracy that soft computing techniques will offer can never be achieved by opting for manual drawings. You have tools to measure the precision, skill and accuracy level of the designs.

Demerits of Soft Computing Technique:

- High production or purchasing cost for new systems
- Time and cost of training the staff which will work on it.
- Need of regular updating of software or operating systems.
- Needs less employment because of soft computing techniques.
- Work can be lost because of the sudden breakdown of computers
- Work is prone to viruses
- Work could be easily "hacked".
- Time taking process to know how to operate or run the soft computing techniques.

1.1.2 Graphical User Interface Software (GUI)

- Designing the visual composition and temporal behavior of a GUI is an important part of software application programming in the area of human-computer interaction.
- Its goal is to enhance the efficiency and ease of use for the underlying logical design of a stored program, a design discipline named usability.
- A GUI (graphical user interface) is a system of interactive visual components for computer software.
- A GUI displays objects that convey information, and represent actions that can be taken by the user. The objects change color, size, or visibility when the user interacts with them.

- GUI objects include icons, cursors, and buttons. These graphical elements are sometimes enhanced with sounds, or visual effects like transparency and drop shadows.
- GUI consists of picture-like items (icons and arrows for example). The main pieces of a GUI are a pointer, icons, windows, menus, scroll bars, and an intuitive input device. Some common GUIs are the ones associated with Microsoft Windows, Mac OSX, Chrome OS, GNOME, KDE, and Android.

1.2 INTRODUCTION, SALIENT FEATURES AND APPLICATIONS OF SOFTWARE'S

- There is many software available which are used in Civil Engineering.
- Due to the Technological Revolution, the number of software serving Civil Engineering needs are increasing manifold.
- So, to list out all this software will be an obtuse thing to do. Here is a list of SOFTWARE which are exorbitantly used by many Civil Engineers all around the world.
- Although, there are various sub disciplines like Transportation Engineering, Structural Design, Surveying, Geo-tech Engineering, Environmental Engineering, Construction planning, we have provided a list of comprehensive list software which are used in civil engineering.

1.2.1 REVIT

- Structural Analysis and Design
- Revit software is specifically built for Building Information Modelling (BIM), empowering design and construction professionals to bring ideas from concept to construction with a co-ordinated and consistent model-based approach.
- It includes the functionality of all of the Revit disciplines (architecture, MEP, and structure) in one unified interface.
- Revit includes tools for architectural design, MEP and structural design, detailing and engineering and construction professionals.
- Contributors across all disciplines use work-sharing to share and save work to the same project.
- Automatically update floor plans, levels, sections and 3D views.
- Use 3D visualizations to see a building before it's built.

1.2.2 ETAB

- Structural Analysis and Design
- ETABS is the ultimate integrated software package for the structural analysis and design of buildings.
- Incorporating 40 years of continuous research and development, this latest ETABS offers unmatched 3D object based modelling and visualization tools, blazingly fast linear and nonlinear analytical power, sophisticated and comprehensive design capabilities for a wide-range of materials, and insightful graphic displays, reports, and schematic drawings that allow users to quickly and easily decipher and understand analysis and design results.
- The System built around a physical object based graphical user interface, powered by targeted new special purpose algorithms for analysis and design, with interfaces for drafting and manufacturing, is redefining standards of integration, productivity and technical innovation.
- The integrated model can include Moment Resisting Frames, Braced Frames, Staggered Truss Systems, Frames with Reduced Beam Sections or Side Plates, Rigid and Flexible Floors, Sloped Roofs, Ramps and Parking Structures, Mezzanine Floors, Multiple Tower Buildings and Stepped Diaphragm Systems with Complex Concrete, Composite or Steel Joist Floor Framing Systems. Solutions to complex problems such as Panel Zone Deformations, Diaphragm Shear Stresses, and Construction Sequence Loading are now at your fingertips.
- Usage and Industry:
  - Design and Analysis.
  - Structural Design Consultancies and firms.

1.2.3 3D Architect Home

- 3D Architect Home Designer enables you to easily draw building project to levels suitable for planning submissions, add detailing and working drawings for building control and visualize it in a detailed 3D model.
- Key Features of 3D Architect Home Designer:
  - Create detailed architectural drawing sets for planning applications.
  - Further enhance 2D drawings with building regulation information.
  - Add detailed hatching to walls and around windows/doors with advanced editor layering option.
  - Draw professional floor plans with wide range of construction tools.
  - Join multiple roof constructions together.
  - Easily modify construction elements with interactive graphics.
  - Define individual areas on walls for applying specific materials.
  - Automatically convert detailed houseplans into fully interactive 3D models.
  - Enhance your home design project with live real-time rendering by adding lighting, shadows and reflections.
- Make direct changes around selected elements in both 2D and 3D with in-line measurements Includes an extensive range of window and door types including bi-folding Import/Export DXF/DWG files.

1.2.4 Build-Master

- Build-Master is a most popular Indian software for RCC building design. It is used by reputed design engineers in more than 250 cities all over India. Developed ingenuously over the years, it is a true "Made in India" product. A complete package for RCC building design, drawing and estimation. It handles all your needs including:
  - Single line floor plan input. Auto numbering of beams and Slabs.
  - Building project defined with floors at different levels.
  - Auto calculation of wall and slab loads on beams.
  - Generation of 3D model, Earthquake/Wind Forces from plans.
  - Loading, BMD, SFD and Deflection Diagrams.
  - RCC design of beams and slabs at all levels, columns and footings.
  - Design schedule data plotted on single line plan for editing.
  - Checks user entered design schedules for adequacy.
  - Schedule and RCC Plan Drawings generated in CAD format.
  - Bar bending schedule of with diameter wise break-up.
  - Quantities of concrete, brickwork, plaster, flooring, formwork.

1.2.5 HEC-RAS

- HEC-RAS is a river modelling software used to compute water surface profiles for steady and unsteady flow models, bridge and culvert roadway crossings, FEMA floodplain encroachments, stream restorations, inline reservoir structures, off-channel storage areas, and much more.
- HEC-RAS provides an easy-to-use, single graphical platform to rapidly create stunning HEC-RAS models and 3D visualizations from a variety of different data sources like AutoCAD, MicroStation, ArcGIS, Google Maps, Bing Maps, and more.

1.2.6 STRAP

- Analysis/Design.
- STRAP is one of the most comprehensive and versatile structural analysis, design and detailing software system available in the market, Today.
- STRAP is extremely used by large construction companies, government agencies, international agencies, engineering firms, Educational and Research organisations worldwide.
- STRAP software's unique design capabilities make it an outstanding Structural design program in the market.

- STRAP comes with design modules for designing and detailing of reinforced concrete structures (frames, slabs, shear walls and foundations), post-tensioned concrete (beams and slabs), structural steel (hot-rolled and cold-formed) and composite members (beams and columns).
- In addition to that STRAP can be used for Automatic bridge analysis and design of water retaining structures.
- STRAP has implemented many design codes like American, European, Canadian, Indian, Australian, South African, Chinese and others.

1.2.7 WaterGEMS

- WaterGEMS provides you with a comprehensive yet easy-to-use decision-support tool for water distribution networks. The software helps improve your knowledge of how infrastructure behaves as a system, how it reacts to operational strategies, and how it should grow as population and demands increase.
- From fire flow and water quality simulations, to criticality and energy cost analysis, WaterGEMS has everything you need in a flexible multiplatform environment.
- WaterGEMS provides numerous software capabilities for:
    - **Intelligent planning for system reliability:** The capability of the water network to adequately serve its customers must be evaluated whenever system growth is anticipated. With WaterGEMS, effectively identify potential problem areas, accommodate service area growth, and plan capital improvements.
    - **Optimized operations for system efficiency:** Realistically modelling the operation of complex water systems can be difficult. With WaterGEMS, model pump accurately, optimize pumping strategies, and plan shutdowns and routine operations to minimize disruption.
    - **Reliable asset renewal decision support for system sustainability:** When it comes time to renew or replace your water infrastructure, the amount of asset-related information you must consider can be overwhelming. WaterGEMS capabilities such as Pipe Renewal Planner makes the task much easier by analyzing and comparing a wide range of variables to prioritize renewal decisions.

1.2.8 Tekla

- Tekla is powerful software developed to meet the needs of the structural engineering workflow and designed to help you automate your repetitive structural calculations.
- Choose from one or more of our regularly updated calculation libraries or write your own, and create professional documentation every time.
- Combine your structural calculations with 2D frame analysis.

1.2.9 ArcGIS

- Geoinformatics
- ArcGIS is a software used in Geographical Information System.
- It is helpful in creating Thematic Maps.
- ArcGIS for Desktop is the key to realizing the advantage of location awareness.
- Collect and manage data, create professional maps, perform traditional and advanced spatial analysis, and solve real problems.
- Make a difference and add tangible value for your organization, your community, and the world.
- Usage and Industry:
    - Surveying, Transportation, mapping.
    - Traffic studies, surveying and Mapping.

1.2.10 QuikGrid

- Geography.
- QuikGrid is a free program for Windows that visualizes data 3D elevation maps.
- If you have ever wanted to visualize your data as a 3D model or heat contour map, QuikGrid is definitely worth checking out.

- This program interprets a set of scattered data points to create maps on a grid; you can choose to either go with a 3D model (using x, y and z points) or with a heat contour map to help visualize your data.

1.2.11 STAAD.Pro

- Structural Analysis and Design
- When it comes to designing and analysis of a structure, STAAD PRO is one such software which is opted by many professionals in construction field.
- STAAD.Pro is the structural engineering professional's choice for steel, concrete, timber, aluminum, and cold-formed steel design of virtually any structure including culverts, petrochemical plants, tunnels, bridges, piles, and much more through its flexible modelling environment, advanced features, and fluent data collaboration.
- Usage and Industry:
  - Design and Analysis.
  - Structural Design Consultancies and firms.

1.2.12 SAFE

- SAFE is the ultimate tool for designing concrete floor and foundation systems. From framing layout all the way through to detail drawing production, SAFE integrates every aspect of the engineering design process in one easy and intuitive environment. SAFE provides unmatched benefits to the engineer with its truly unique combination of power, comprehensive capabilities, and ease-of-use.
- Laying out models is quick and efficient with the sophisticated drawing tools, or use one of the import options to bring in data from CAD, spreadsheet, or database programs. Slabs or foundations can be of any shape, and can include edges shaped with circular and spline curves.
- Post-tensioning may be included in both slabs and beams to balance a percentage of the self-weight. Suspended slabs can include flat, two-way, waffle, and ribbed framing systems. Models can have columns, braces, walls, and ramps connected from the floors above and below. Walls can be modelled as either straight or curved.
- Mats and foundations can include nonl-inear uplift from the soil springs, and a non-linear cracked analysis is available for slabs. Generating pattern surface loads is easily done by SAFE with an automated option. Design strips can be generated by SAFE or drawn in a completely arbitrary manner by the user, with complete control provided for locating and sizing the calculated reinforcement. Finite element design without strips is also available and useful for slabs with complex geometries.
- Comprehensive and customizable reports are available for all analysis and design results. Detailed plans, sections, elevations, schedules, and tables may be generated, viewed, and printed from within SAFE or exported to CAD packages.
- SAFE provides an immensely capable yet easy-to-use program for structural designers, provideing the only tool necessary for the modelling, analysis, design, and detailing of concrete slab systems and foundations.

1.2.13 RISA Connection

- RISA Connection is a useful program designed to assist engineers in finishing steel structures by creating welds and connections between the elements. The program uses a visual approach but you can also enter data by editing the design spreadsheets. You can use the program to inspect the design in a 3D view and to modify the dimensions in the 2D panel. Both views can be exported and printed in order to use them in your project documentation.
- RISA Connection takes the complexity out of calculations by listing explicitly the calculations and the code check values. Review your results in an easy-to-read report or inspect the connection in a 2D shop drawing-style drawing.
- RISA Connection allows you to quickly build steel connections using a quick connection dialog box with pictures and descriptions.
- Review your connection in an interactive 2-dimensional picture or view it fully rendered in 3-D. The rendered view allows you to rotate 360 degrees to see the entire connection.

1.2.14 Civil 3D

- Civil 3D software is a civil engineering design and documentation solution that supports Building Information Modelling (BIM) workflows on civil infrastructure projects, including roads and highways, rail, site development, airports, and water.
- Civil 3D helps civil infrastructure professionals improve project delivery, reduce risk, errors, and omissions, and respond faster to project changes.
- Additionally, users can streamline time-consuming tasks such as corridor design, (road and rail), intersection design, parcel layout, site grading, and piping design with specific tools and customizable design standards.

1.2.15 Site 3D

- Site 3D is a fully featured software product for the engineering design of highways, drainage and residential developments.
- Quickly and accurately design whole sites including roads, junctions, roundabouts, footways, surface interfacing, contours, volumes, drainage, housing and more.
- Site 3D is a powerful software tool for the 3D engineering design of road schemes and infrastructure.

1.2.16 SkyCiv Structural 3D

- SkyCiv Structural 3D is a cloud-based Structural Analysis and Design software that allows users to model and solve complex 3D structures.
- It is completely online; requiring no installation to use.
- SkyCiv Structural 3D offers the very best in online structural analysis and integrated design checks.
- Structural 3D is powerful analysis software that includes:
  - Reactions, Bending, Shear, Axial and Torsion Forces.
  - Stress, Deflection and Plate Analysis.
  - AISC 360, EN 3, AS 4100, CSA S16 design checks.
  - Fully Integrated Section Builder (with custom and database shapes).
  - P-Delta (Non-linear) and Linear Analysis.
  - Buckling Analysis - Full 3D Renderer with Colored Contouring.
  - PDF Analysis and Design Reporting.
  - Integrated design checks including AISC, Eurocode and Australian Standards.
  - Import Load Combinations for US, UK, Aus, Canada and Europe.
- With the Structural 3D Package, the following tools are included with your membership:
  - SkyCiv Structural 3D
  - SkyCiv Beam
  - SkyCiv Section Builder
  - SkyCiv Truss
  - SkyCiv Frame
  - SkyCiv Shaft Show Less

1.2.17 SAP 2000

- Structural Analysis and Design.
- SAP2000 is another software available for design and analysis of structures.
- It is also used in many structural firms. SAP2000 has a wide selection of templates for quickly starting a new model.
- SAP2000 includes parametric templates for the following types of structures: Simple Beams, 3D Trusses, 3D Frames, Storage Vessels, Staircases, Dam Structures, and Pipes.
- View and manipulate analytical and physical models with great precision. Easily define custom views and developed elevations to view and manipulate complex geometry with ease.

- SAP2000 automatically create joints at structural object intersections or internal joints when meshing structural objects. Joint co-ordinates and information may be displayed on screen in the model window or in tabular format.
- Usage and Industry:
  - Design and Analysis.
  - Structural Design Consultancies and firms.

1.2.18 MIDAS

- PSC Girder Design
- MIDAS Civil is state of the art engineering software that set a new standard for the design or bridges and civil structures.
- It features a distinctively user friendly interface and optimal design solution functions that can account for construction stages and time dependent properties.
- Its highly developed modelling and analysis functions enable engineers to overcome common challenges and inefficiencies of finite element analysis.
- With MIDAS Civil, you will be able to create high quality designs with unprecedented levels of efficiency and accuracy.

1.2.19 LUSAS

- LUSAS Civil and Structural is a world-leading finite element analysis software application for the analysis, design and assessment of all types of structures.
- Whether you need to carry out a straightforward linear static analysis of a simple structural frame, a dynamic analysis of a masonry or concrete building, or a detailed geometrically nonlinear staged erection analysis of a cable stayed structure, LUSAS Civil and Structural provides all the facilities you need.
- We develop, market and support a range of engineering analysis software products all based upon the LUSAS finite element analysis system.
- LUSAS software products can solve all types of linear and nonlinear stress, dynamics, composite and thermal engineering analysis problem, across a range of engineering industries.
- LUSAS consultancy services solve a wide range of analysis problems for clients in all engineering industries.

1.2.20 BricsCAD

- BricsCAD is a software application for computer-aided design (CAD), developed by Bricsys nv.
- BricsCAD is an all-in-one computer-aided design (CAD) software that combines 3D modelling and 2D CAD for .dwg. It's a less popular platform that provides all .dwg 2D CAD features while bringing the convenience of the powerful 3D modelling and time-saving tools. The program is expertly crafted with engineers in mind to support efficient execution of all projects.
- The application provides intuitive modelling and drawing solutions and is available in 3 editions namely; Classic, Pro, and platinum. It's also open to 3rd party developers making it an ideal solution for users looking to scale past the AutoCAD horizon.
- Some of its key features include an intelligent quad cursor, settings manager, drawing explorer, dockable panel, layers and content browser, web map services, dynamic arrays, annotative scales, dynamic dimensions, and 3D modelling.

1.2.21 Estimate Master

- Estimate Master is a fully featured Construction Management Software designed to serve Enterprises, Startups. Estimate Master provides end-to-end solutions designed for Web App.
- This online Construction Management system offers Accounting Integration, Audit Trail, Historical Database, Project Management, Reporting/Analytics at one place.
- A general purpose estimating tool for any business. Adaptable to different types of businesses by virtue of an open database, customizable estimate fields and customizable cost basis categories.
- Optionally works with QuickBooks. Produces formal customer bids. Can also produce subcontractor work orders from estimates. Estimate Master works with Schedule Master will which turn your completed estimates into schedules.

1.2.22 ProEst

- ProEst is a cloud-based construction estimating solution that offers automated features for construction companies of all sizes.
- ProEst enables construction teams to monitor and manage centralized estimates, takeoffs, reports and contracts, and all information can be accessed and shared 24/7 from any internet-connected mobile device.
- Dashboards offer a look at relevant KPIs, so users can measure, evaluate and control business performance throughout the project lifecycle, and construction databases provide current costing and productivity data.
- ProEst also offers visibility into the future by tracking pending projects, updating estimate pipelines and creating accurate short and long-term business forecasts.
- ProEst is available on an annual subscription basis that supports unlimited users.

1.2.23 WinEst

- With a proven record of continually developing the art and science of cost-estimating technology, WinEst serves the quantity takeoff and cost estimating needs of building construction clients, large and small, all around the globe.
- Trimble WinEst serves the complete range of industry niches and customers - from regional single-office builders to global general contractors, facility owners, specialty contractors, public/government agencies and more.
- WinEst is an on-premise database-driven estimating solution that helps estimators in businesses across various industries to manage and integrate project estimates using cost-knowledge management.
- Key features include construction takeoff, estimating, audit trails and reporting.
- Users can set-up project estimates from scratch or with a project-type template that follows corporate standards for general conditions and indirect costs. The solution's database structure allows tracking of the audit trail for changes to the estimate. The entire estimate can be filtered to make a final review of the alerts. It also allows users to come up with what-if scenarios by creating alternatives like landscaping, a surveillance system, a sound system or a flagpole.
- WinEst provides organizations various methods to code their estimating data by CSI, Masterformat or Uniformat. The solution seamlessly integrates with various accounting, project scheduling, project management and CAD solutions.
- Support is offered via phone, email and through other online resources.

1.2.24 Clear Estimate

- Clear Estimates is an easy-to-use, residential remodelling estimating program that allows you to quickly determine a project's cost, manage and customize a part database, maintain a customer and project library, produce several types of reports and proposals, process change orders, integrate with Quickbooks, and access up to date pricing specific to your area.
- Clear Estimates is a construction estimating system for remodelers and contractors which allows them to create estimates, produce bid requests, draft proposals, manage customers and more.
- It integrates with QuickBooks and allows users to export project estimates and customer information into the system. It also relies on the RemodelMAX database for material and labor cost information.
- While originally intended for remodelers, the system can be customized to meet the needs of most trades, such as, homebuilders, design/build firms, finishing contractors, roofing and siding contractors and other specialty contractors.

1.2.25 Procore

- Procore manages your projects, resources and financials from project planning to closeout. The platform connects every project contributor with the owner and general or specialty contractor. The centralized dashboard allows managers to handle project details, schedule tasks and view progress.

- It enables users to communicate across disparate teams, making it easier to work together by eliminating silos and establishing a single source of information.
- Procore gives your team access to everything they need to get their job done.
- It is available on monthly subscriptions and support is extended via phone, email, knowledgebase and other online measures.

1.2.26 Builder Trend

- Builder trend is a construction management system for home builders, remodelers, specialty contractors and general contractors.
- It combines project scheduling, project management, financial management, customer management and service management in a single suite.
- As a cloud-based platform, this system can be accessed online with a computer or mobile device.
- Builder trend provides pre-sale tools including a built-in customer relationship management (CRM) system, bids and proposals. Project management tools include scheduling, budgeting, time sheets and more.
- Customer management tools include change order and selection management, warranty requests and payment processing.
- This system also integrates with other solutions such as QuickBooks, Xero and various dedicated estimating and takeoff tools.
- Buildertrend provides each company with an account manager to provide training and support.

1.2.27 Building Management System (BMS)

- Building Management System (BMS), is software that allows users to monitor and control equipment within a building.
- A well-structured and user friendly BMS ensures easy operation of buildings and allows insight into all building systems.
- Optergy's built-in applications enable contextual displays, smart alarming and analysis. Optergy's software allows integration with BACnet and Modbus protocols, the industry's standard for HVAC equipment.
- Using a common web browser and a mobile device, users can stay connected and have full control over the building assets.

1.2.28 Plant Design and Management System (PDMS)

- Modelling.
- PDMS (Plant Design Management System) as it is known in the 3D CAD industry, is a customizable, multi-user and multi-discipline, engineer controlled design software package for engineering, design and construction projects in offshore and on shore.
- PDMS is also used for engineering, design and construction projects, chemical and process plants, water treatment plants, power plants, paper and pulp manufacturing. PDMS also has a marine application as well.
- Equipment and Piping Modelling, Preparation of 3D Piping Layout with ducting layout drawings, reference to PandID and 2D Piping Layout; Creation of equipment models using
- Existing paper drawings or static cad files, 3d modelling of equipment, piping, structures, instrumentation and other facilities and extraction of detailed construction drawing and documents.
- Structural modelling, HVAC, cable tray and cabling system.
- Isometric generation and spooler: preparation of isometric drawing of piping route for bill of Quantity and creation of 2d draft and isometric drawings using PDMS.

1.2.29 Building Information Modelling (BIM)

- Building Information Modelling can be termed as one of the 21st century's greatest innovation helping big infrastructure projects.
- BIM (Building Information Modelling) is an intelligent 3D model-based process that gives architecture, engineering, and construction (AEC) professionals the insight and tools to more efficiently plan, design, construct, and manage buildings and infrastructure.

1.2.30 Primavera Pro

- Project Management.
- Primavera is a construction Planning and Scheduling software.
- It is opted by almost greater than 70 percent of all Planning and scheduling professionals.
- MS Office also offering similar software by name MS PROJECT.
- Primavera offers best-in-class solutions focused on the mission critical PPM requirements of key vertical industries including engineering and construction, public sector, aerospace and defense, utilities, oil and gas, manufacturing and high tech, and IT and services.
- Primavera PPM products, together with Oracle's project financials, human resources, supply chain management, product lifecycle management, business intelligence, and infrastructure software are expected to provide the first, comprehensive Enterprise Project Portfolio Management solution.
- This solution is expected to help companies optimize resources and the supply chain, reduce costs, manage changes, meet delivery dates, and ultimately make better decisions, all by using real-time data.
- Usage and Industry:
  - Design and Analysis.
  - Structural Design Consultancies and firms, Project planning and consultants.

1.2.31 Microsoft Project (MSP)

- Project Management.
- Microsoft project is a project management software product.
- It is developed and sold by Microsoft.
- It is designed to assist a project manager developing a plan, managing the budget and analyzing workload.
- As resources are assigned to tasks and assignment work estimated, the program calculates the cost, equal to the work times the rate.

Important Points

- Soft computing is the use of approximate calculations to provide imprecise but usable solutions to complex computational problems.
- Soft computing is sometimes referred to as computational intelligence. Soft computing provides an approach to problem-solving using means other than computers.
- The principal constituents of Soft Computing (SC) are Fuzzy Logic (FL), Evolutionary Computation (EC), Machine Learning (ML) and Probabilistic Reasoning (PR), with the latter subsuming belief networks and parts of learning theory.
- Soft computing is an important branch of computational intelligence, where fuzzy logic, probability theory, neural networks, and genetic algorithms are synergistically used to mimic the reasoning and decision making of a human. Approaches that are useful in the development of an intelligent machine are introduced.
- Soft computing, as opposed to traditional computing, deals with approximate models and gives solutions to complex real-life problems. Unlike hard computing, soft computing is tolerant of imprecision, uncertainty, partial truth, and approximations. In effect, the role model for soft computing is the human mind.
- The applications of soft computing approach have proved two main advantages:
  (1) it made solving nonlinear problems, in which mathematical models are not available, possible and
  (2) it introduced the human knowledge such as cognition, recognition, understanding, learning, and others into the fields of computing.
- A GUI (Graphical User Interface) is a system of interactive visual components for computer software. A GUI displays objects that convey information, and represent actions that can be taken by the user. The objects change color, size, or visibility when the user interacts with them.

- GUI consists of picture-like items (icons and arrows for example). The main pieces of a GUI are a pointer, icons, windows, menus, scroll bars, and an intuitive input device. Some common GUIs are the ones associated with Microsoft Windows, Mac OSX, Chrome OS, GNOME, KDE, and Android.
- Autodesk Revit Architecture is used by architects and other building professionals to help reduce risk, obtain insight into how buildings will perform before construction begins, develop better quality designs, and improve project delivery.
- A good way to look at the benefits of each tool is AutoCAD is great for 2D drawing, where only precise line work is needed, such as elevation detail drawings. Revit is great for modelling, generating cost schedules, collaboration and change management. In the industry of design and construction, competition is fierce.
- ETABS is an engineering software product that caters to multi-story building analysis and design. Modelling tools and templates, code-based load prescriptions, analysis methods and solution techniques, all coordinate with the grid-like geometry unique to this class of structure.
- Generally, ETABS is used for the analysis of concrete shear walls and concrete moment frames. Once we are able to limit the drift, we can output the forces from ETABS into a spreadsheet for design. We design the piers of the shears walls for flexure and axial loads using PCA Column. We design the moment frames using PCA. ETABS is a highly efficient analysis and design program developed especially for building systems. It is loaded with an integrated system with an ability to handle the largest and most complex building models and configurations.
- 3D Architect Home Designer Pro enables you to easily draw building project to levels suitable for planning submissions, add detailing and working drawings for building control and visualise it in a detailed 3D model.
- BuildMaster is an application release automation tool, designed by the software development team, Inedo. It combines build management and ARCHITECTURE capabilities to manage and automate processes primarily related to continuous integration, database change scripts, and production deployments, overall releasing applications reliably. The tool is browser-based and able to be used "out-of-the-box". Its feature set and scope puts it in line with the DevOps movement, and is marketed as "more than a release automation tool. It brings together the people, processes, and practices that allow teams to deliver software rapidly, reliably, and responsibly." It's a tool that embodies incremental DevOps adoption.
- BuildMaster is configured entirely through its UI, as opposed to scripts or XML-based configuration files. Although the primary web application is Windows-only, BuildMaster orchestrates Windows or Linux-based servers to perform various build-release-deploy actions. BuildMaster also has a tight integration with PowerShell and is often used along with other DevOps tools.
- HEC-RAS is a computer program for modelling water flowing through systems of open channels and computing water surface profiles. HEC-RAS finds particular commercial application in floodplain management and [flood insurance] studies to evaluate floodway encroachments.
- The Hydrologic Engineering Center's (HEC) River Analysis System (HEC-RAS) software allows the user to perform one-dimensional (1D) steady and 1D and two-dimensional (2D) unsteady flow river hydraulics calculations. HEC-RAS is an integrated system of software.
- STRAP (STRuctural Analysis Programs) is a Windows-based suite of finite element static and dynamic analysis programs for buildings, bridges and other structures.
- STRAP is one of the most comprehensive and versatile structural analysis and design software systems available on the market today.
- Water GEMS is a hydraulic and water quality modelling solution for water distribution systems with advanced interoperability, geospatial model-building, optimization, and asset management tools.
- Water GEMS is a superset of Water CAD, so you will get everything you have in Water CAD plus more with Water GEMS. ... The user interface is identical with the exception of the tools that are not available in Water CAD. There are two main differences between Water CAD and Water GEMS.

- Tekla Structures is a building information modelling software able to model structures that incorporate different kinds of building materials, including steel, concrete, timber and glass. ... Tekla Structures was formerly known as Xsteel (X as in X Window System, the foundation of the Unix GUI).
- ArcGIS Online is a cloud-based mapping, analysis, and data storage system hosted by Esri that can be used to create, share, and manage maps, scenes, layers, apps, and other geographic content. ... ArcGIS for Developers is a website that contains all of the resources you need to build and deploy ArcGIS applications.
- Esri's ArcGIS Desktop is a suite of integrated applications, including ArcMap, ArcCatalog, and ArcToolbox. By using these applications and interfaces in unison, you can perform any GIS task, including mapping, geographic analysis, data editing and compilation, data management, visualization, and geoprocessing.
- QUIKGRID is a program which will read in a set of scattered data points (x, y, z) which represents a surface. The program will generate a grid from this data and then display the surface as a contour map, or as a 3d representation. The left mouse button may be used to zoom in and pan about the display. The right mouse button may be used to display information about the generated grid coordinates and data points.
- STAAD or (STAAD.Pro) is a structural analysis and design software application originally developed by Research Engineers International in 1997. In late 2005, Research Engineers International was bought by Bentley Systems.
- Staad Pro is structural software which can be used for Analysis of structural members of the structure. It's mostly used for exicut the structural elements like columns, beams, walls, slabs on basis of loads moments, shear forces, torsional effects, bending moments, deflection etc.
- Staad Pro. Staad is a design tool solely meant for structural engineers, focusing on structural analysis and structural design. Civil engineers may learn how to use Staad Pro, but they can use it if they have prospects of working as structural engineers.
- RISA has been developing world-class structural design software for over 30 years. ... The seamless integration of our product suite creates a powerful, versatile and intuitive structural design environment, ready to tackle almost any design challenge.
- Regional Integrated Sciences and Assessments (US NOAA) RISA. Rapid Interactive Structural Analysis (structural engineering software) RISA. Radioactive Iodinated Serum Albumin.
- The educational version of the software RISA-2D, developed by RISA Technologies for the textbook Fundamentals of Structural Analysis, is an interactive computer program for the analysis of two-dimensional structures such as continuous beams, trusses, and frames.
- AutoCAD Civil 3D software is a design and documentation solution for civil engineering that supports building information modelling (BIM) workflows. By learning to use AutoCAD Civil 3D, you can improve project performance, maintain consistent data, follow standard processes, and respond faster to change.
- Autodesk Civil 3D software is a design and documentation solution for civil engineering that supports building information modelling (BIM) workflows. By learning to use Autodesk Civil 3D, you can improve project performance, maintain consistent data, follow standard processes, and respond faster to change.
- SkyCiv is a company that provides users with robust yet easy-to-use online structural analysis software. SkyCiv cloud engineering software is easy-to-use and user friendly. Plus, their pricing schedules are affordable compared to conventional engineering software.
- SkyCiv Structural 3D is our most powerful structural analysis software on the cloud. This 3D frame analysis software allows you to model, analyse and design a wide range of structures.
- The 100% cloud-based engineering Software allows users to access their programs and files from a web browser, from anywhere, anytime! No need to install or update.
- Different from traditional software that are clunky and difficult to use, SkyCiv is by far the most intuitive and easy-to-use structural engineering software for civil and structural engineers.
- SAP2000 is general-purpose civil-engineering software ideal for the analysis and design of any type of structural system. ... Created by engineers for effective engineering, SAP2000 is the ideal software tool for users of any experience level, designing any structural system.

- SAP2000 is intended for use on civil structures such as dams, communication towers, stadiums, industrial plants and buildings. Once you go thorough this course, you can use this software for structural analysis and design of any kind of civil engineering structure.
- MIDAS Civil is state of the art engineering software that set a new standard for the design or bridges and civil structures. It features a distinctively user friendly interface and optimal design solution functions that can account for construction stages and time dependent properties.
- LUSAS software consists of a Windows-based Modeller, used for model building and viewing of results, and a Solver for carrying out an analysis. Four commercial application products cater for the following industries: Civil & Structural - for civil, structural, nuclear, seismic, geotechnical and offshore engineering.
- BricsCAD is a software application for computer-aided design, developed by Bricsys nv. The company was founded in 2002 by Erik de Keyser, a longtime CAD entrepreneur.
- BricsCAD is a software application for computer-aided design (CAD), developed by Bricsys nv. The company was founded in 2002 by Erik de Keyser,[1] a longtime CAD entrepreneur. In 2011 Bricsys acquired the intellectual property rights from Ledas for constraints-based parametric design tools, permitting the development of applications in the areas of direct modelling and assembly design.
- Estimate Master is a general estimating tool for any business. Adaptable to different types of businesses by virtue of an open database and customizable estimate fields and cost basis categories.
- ProEst is a cloud-based construction estimating solution that offers automated features for construction companies of all sizes.
- PlanSwift is an on-screen digitizing and takeoff system that is suitable for both commercial and residential construction and subcontractors in a number of industries, such as electrical, heavy/highway, plumbing, and custom and production home builders.
- WinEst is an on premise database-driven estimating solution that helps estimators in businesses across various industries to manage and integrate project estimates using cost-knowledge management. WinEst provides organizations various methods to code their estimating data by CSI, Master format.
- Clear Estimates is a construction estimating system for remodelers and contractors which allows them to create estimates, produce bid requests, draft proposals, manage customers and more as an online, cloud-based tool.
- ProCore is the construction industry's leading construction management software that fully integrates with ProEst estimating software. Construction companies can use ProCore to manage all phases and aspects of projects, from pre-development and bidding to project completion.
- Procore is your complete online construction management software. Procore provides cloud-based applications to help construction professionals build quality projects on time and within budget.
- Buildertrend is a construction management system for home builders, remodelers, specialty contractors and general contractors. It combines project scheduling, project management, financial management, customer management and service management in a single suite.
- Building Management System (BMS), is software that allows users to monitor and control equipment within a building. A well structured and user friendly BMS ensures easy operation of buildings and allows insight into all building systems.
- BMS stands for Building Management System whereas HVAC stands for Heating Ventilation and Air Conditioning. HVAC system is responsible for providing desired parameters of temperature and humidity. Normally now a days, HVAC system is installed with B.M.S. for energy efficiency.
- PDMS as it is known in the 3D CAD industry, is a customizable, multi-user and multi-discipline, engineer controlled design software package for engineering, design and construction projects in offshore and onshore.

- Building Information Modelling (BIM) is an intelligent 3D model-based process that gives architecture, engineering, and construction (AEC) professionals the insight and tools to more efficiently plan, design, construct, and manage buildings and infrastructure.
- "BIM is an intelligent model-based process that helps make design, engineering, project and operational information accurate, accessible and actionable for buildings and infrastructure," according to Bond, PR manager for AEC and infrastructure at Autodesk. This data is key for a building information modelling project.
- Primavera is an enterprise project portfolio management software. It includes project management, scheduling, risk analysis, opportunity management, resource management, collaboration and control capabilities, and integrates with other enterprise software such as Oracle and SAP's ERP systems.
- Primavera P6 is a software used by owned by Oracle. It is used by project management professionals and also can be linked with ERP systems. It handles multiple projects and can hold up to 100,000 activities with unlimited resources and an unlimited number of target plans.
- Primavera project management software was created as an enterprise software from the beginning whereas the MS-Project software was earlier created as a desktop application and then it was migrated to project server and the enterprise capabilities were added to the Microsoft project software.
- Managed service provider (MSP) software is an IT service platform that manages a predefined set of services for clients. Working within a number of industries, such as B2B integration, information services, supply chain management, transportation, utilities, and media, MSP software works as a virtual assistant.

Multiple Choice Questions

1. Which one is not a soft computing technique ?
   (a) Fuzzy logic (b) Machine learning
   (c) Evolutionary computation (d) Traditional computing
2. Use of software in Civil Engineering problems ......
   (a) saves time (b) increase accuracy
   (c) data can be stored and restored (d) all of the above
3. "GUI" means ....
   (a) golden unit of internal software (b) graphical user interface
   (c) graphical user intelligently (d) golden user interface
4. Microsoft windows, Android, Mac OSX are fome of the ......
   (a) computers (b) output devices
   (c) GUI (d) softwares
5. "Revit" software is meant for ......
   (a) revision of vision interface transformation (b) revise instantly
   (c) revise immediately (d) random vision instantly
6. Design of buildings with plan, elevation, sections can be done by ......
   (a) Tally (b) CAD
   (c) C^{++} (d) none of these
7. Software which is used for modelling of building, cost schedule, estimating of project is ......
   (a) Tally (b) Excel
   (c) Revit (d) Oracle
8. Software used for analysis and design of tall R.C.C. structure is ......
   (a) Etab (b) Cad
   (c) Revit (d) none of these
9. "PCA" means ......
   (a) principal component analysis (b) principal computational analysis
   (c) point column analysis (d) none of these

10. Audesk company has developed a software for preparation of building drawings and drawings of machine components ......
    (a) Archicad (b) Lumion
    (c) Cedreo (d) Autocad
11. Floor planner is a software used for drawing ......
    (a) elevations (b) sections
    (c) floor plans (d) 3-d model
12. "Build master" is used for ......
    (a) R.C.C. design (b) steel structural design
    (c) estimating (d) none of these
13. "HEC-RAS" is a software and it is called ......
    (a) hydraulic engineering center's river analysis system
    (b) hydrographic estimation centre of random access software
    (c) hydro evaluation centre for river and small dam
    (d) none of these
14. "STRAP" is a software used for ......
    (a) soil analysis (b) structural analysis
    (c) canal analysis (d) none of these
15. For hydraulic and water modelling solutions which software is used.
    (a) STRAP (b) Antocad
    (c) Water Gems (d) ETab
16. "Tekla" structures is a software for ......
    (a) building information modelling (b) preparation of plans and elevation
    (c) preparation of models of building (d) none of these
17. A cloud base mapping, analysis and data storage system hosted by Esri is ......
    (a) Tekla (b) ArcGis
    (c) Arch.Desktop (d) Water Gems
18. Software program used to set of scattered data points (x, y, z) is ......
    (a) STAAD (b) Quickgrid
    (c) ArcGis (d) Tekla
19. Structural engineers uses which software for analysis and design of structure ?
    (a) Staad-Pro (b) RisA
    (c) Water Gems (d) Quickgrid
20. Structural analysis software on cloud is ....
    (a) Skyciv (b) RisA-ZD
    (c) Civil-3D (d) Stadd pro
21. "SAP-2000" is a software used for ......
    (a) Effective engineering (b) For analysis is of dam
    (c) For analysis of tower (d) All of these
22. Which of these are softwares used in Civil Engineering only ?
    (a) Autocad, E-Tab, Staad-pro (b) Autocad, Talley, WinEst
    (c) Sap-2000, Tally, Excel (d) None of these

Answers

1. (d)	2. (d)	3. (b)	4. (c)	5. (b)	6. (b)	7. (c)	8. (a)	9. (a)	10. (d)
11. (c)	12. (c)	13. (a)	14. (c)	15. (c)	16. (a)	17. (b)	18. (b)	19. (a)	20. (a)
21. (d)	22. (a)								

❑❑❑

Chapter 2

Recent Construction Materials

Weightage of Marks = 14, Teaching Hours = 10

Syllabus

2.1 Building Materials

- Artificial sand, Sensi tile, carbon fibre, Bricks made-up of cigarette butts, 3D printed bricks, Translucent wood, laminated timber, 3D Tiles

2.2 Road Materials

- Geo-synthetics, Noise-reducing asphalt, Porous Pavement, Plastic Roads, solar roads, Anti Icing Roads, Piezoelectric roads

2.3 Concrete Materials

- Portland Pozzolana Cement, Portland Slag Cement
- New admixtures — Masterglanium, Polycarboxylic Ether
- Self Healing Concrete, Fibre-Reinforced Concrete, High Strength concrete, High Performance Concrete, Nano concrete, Light transmitting concrete.

2.4 Sustainable Materials

- Ground Granulated Glass Blast-furnace Slag (GGBS) concrete, Aero-gel insulation, Cooling bricks, Green concrete, Timbercrete, Ferrock

About this Chapter

At the end of this chapter, students will be able to:

- List the advanced construction materials for given construction project.
- Identify the properties of given advanced material.
- Enumerate the applications of given form of material/s in civil engineering.
- Identify the situations to use the given material.
- Justify the use of given material for given site condition.

2.1 BUILDING MATERIALS

- Introduction and implementation of super performing building materials and techniques all in terms of energy saving efficiency of the material, cost efficiency, application feasibility, availability, vernacular characteristics, life span etc.
- Smart structures and material technologies are a tool for sharing the knowledge of how various building materials can significantly increase production and profit using advanced communication, collaboration and management technologies.

2.1.1 Artificial Sand

- Artificial Sand is produced by crushing stones obtained from rock. The need for artificial sand arises due to non-availability of good river sand. The natural river sand requires transportation from long distances, whereas artificial sand can be produced in local area.
- Machineries required for artificial sand are : Jaw crushers, (ii) Cone crushers and (iii) Roll crusher. The grains of artificial sand that is cubical and angular in shape produce good quality of mortar and concrete.

- Natural sand are weathered and worn out particles of rocks and are of various grades or size depending on the accounting of wearing. The main natural and cheapest resource of sand is river. Dams are constructed on every river hence these resources are erasing very fast. Now-a-day's good sand is not readily available, it should be transported from long distance. Those resources are also exhausting very rapidly. So it is a need of the time to find some substitute to natural river sand.
- The artificial sand produced by proper machines can be a better substitute to river sand. The sand should be sharp, clean and course.
- The grains should be of durable material. The grain sizes must be such that it should give minimum voids.
- The presence of clay and silt retards the setting of the cement and makes the mortar weaker and the walls or the slab leaks and holds dampness. The sand in the mortar does not add any strength but it is used as an adulterant for economy and with the same it prevents the shrinkage and cracking of mortar in setting.
- The sand must be of proper gradation (it should have particles from 150 microns to 4.75 mm in proper proportion) When fine particles are in proper proportion, the sand will have less voids.
- The cement required will be less when there will be less void in sand. Such sand will be more economical.

Properties of Artificial Sand

- The sand particles have a higher crushing strength than natural sand.
- The surface texture of the particles should be smooth.
- There should not be any organic impurities. It should be free of clay and silt.
- Grains should have cubical and angular shape.
- It should be well-graded having sizes of particles from 15011 to 4.75 mm.

Advantages of Artificial Sand

- As there is a ban imposed by government on dredging of natural sand, this is a better substitute to river sand.
- Natural sand takes millions of years to form whereas the artificial sand is produced in short period of time.
- Transportation cost is reduced as it can be produced within a city.
- Sand of required particle size can be produced as per requirement of mix design.

2.1.2 Sensi Tile

- Think about if you walk across your kitchen floor to get something from the refrigerator, the floor twinkles with a lighted path that guides your way through the dark room. This is now possible if you had Sensi tiles.
- In this type of tiles concrete acrylic fiber-optic channels that transfer light from one point to another are embedded in it.
- As person move across Terrazzo's surface, the light channels flicker with a randomized, twinkling effect.
- Tiles are available for use as flooring, in bathrooms and even ceilings, so you can have twinkling lights follow you all over the house.

2.1.3 Carbon Fibre

- Carbon fiber is a material which is extremely strong and light weighted.
- The carbon fiber material is 5 times as strong as steel, two times as stiff, yet weighs about two-thirds less.
- These fibers are made-up of carbon strands that are thinner than a human hair.

- The carbon strands can be woven together, like cloth, and then that can be molded to any shape you might want.
- In addition, carbon fibers are strong as well as flexible, so it's the perfect material for construction projects in areas with that are exposed to hurricanes and tornados.

2.1.4 Bricks made-up of Cigarette Butts

- Cigarette butt bricks are produced by blending cigarette butts with mixture of materials used for brick production. This is an innovative and practical solution for the pollution problems caused by cigarette butt disposal.
- The percentage of cigarette butt addition is based on the purpose for which bricks are being used. This is because a high percentage of cigarette butts would lead to the reduction of brick dry density and compressive strength.
- Not only does the addition of cigarette butt cut brick production costs, but also results in lighter bricks, with better insulation properties, reducing household heating and cooling costs.
- This innovative method of brick production makes bricks more environmentally friendly product because it leads to decreased dumped cigarette buts and reduces energy used during brick production process.
- Cigarette butts are more than just pesky litter, their filters contain heavy metals like arsenic, chromium, nickel and cadmium and since they take many years to break down, that gives those metals plenty of time to leach into the soil and water.

Physical Properties

The following properties are valid for bricks with 10% cigarette butt content:

- Density of cigarette butt Bricks is less than that of conventional bricks without cigarette butts.
- Compressive strength of cigarette butt bricks is less than the conventional bricks by more than 85%.
- Water absorption increases almost linearly with the increase of cigarette butt content.
- Cigarette butt bricks have low porosity and shrinkage.
- Bricks are light and have good insulation capability.
- The percentage of cigarette butt addition is specified based on the type of project or works in which cigarette butt bricks are used. This is because different amount of cigarette butt in bricks modify brick properties.
- Tests have been conducted on different replacement percentages like 1.5, 2.5, 7.5, and 10 percent. Each replacement quantity results in different physical properties. As the quantity of cigarette butts is increased, its detrimental effects on the bricks would be more obvious.

Advantages

- It gets that waste out of the environment.
- It makes bricks cheaper.
- Less energy-intensive bricks can be produced.
- The energy needed to fire bricks can be cut by up to 58%.
- Lighter and better insulator bricks are produced; meaning they could help cut household cooling and heating demands.
- When the cigarette butts are fired in the bricks, the heavy metals and other pollutants are trapped and immobilized in the solid block so they cannot leach.
- Bricks shrinkage, porosity, and thermal properties are improved.
- Cigarette butts can be placed in bricks without any fear of leaching or contamination.
- Cigarette butt bricks are an alternative to conventional bricks.

2.1.5 3D Printed Bricks

- 3D Printed Bricks are be the building blocks of modern, sustainable architecture people have built homes by shaping mud into blocks, drying them in the sun, and stacking them.
- Using clay and sand, 3D print bricks are designed blocks that can cool down rooms and even stabilize buildings when earthquakes strike.
- They are also working with unconventional materials such as salt and sawdust—techniques that could one day allow waste to be transformed into dwellings.
- The Cool Brick is a 3D-printed ceramic object about six inches long, shaped like an elongated hexagon.
- In addition to being very porous, the brick resembles a small lattice with larger holes that air can blow through.
- When water is dripped onto the wall with a plumbing system or splashed on by hand, it is absorbed into the ceramic and then evaporates, cooling the air.
- Breezes blowing through the holes bring cool, humid air into the room.

2.1.6 Translucent Wood

- Translucent wood is one of the innovative construction materials that hold the potential to lift the construction sector to a whole new level. Not only is it environmentally-friendly, but also a great alternative to plastic and glass.
- Translucent wood is produced by treating and compressing wood strips. In the production process, lignin is replaced by polymers to make the wood translucent.
- As far as translucent wood's strength is concerned, it has the same strength of lumber whilst being far lighter. Finally, translucent wood could be utilized in home construction and bring more light into the house, consequently reducing the need for artificial light which can consume a lot of power
- The technique that is used to produce translucent wood is similar to chemical pulping:
  - Place wood blocks in boiled water and sodium hydroxide for two hours. This would leach out polymer called lignin (gives wood its color) in the wood.
  - Then, the wood is soaked with an epoxy which not only strengthens the wood but also turns it clear.

Properties

Properties of translucent wood are strongly dependent on wood properties:

- Low density compared to glass (from 1200 kg/m^3).
- High optical transmittance.
- Low thermal conductivity.
- Outstanding toughness.

Applications

- Smart buildings with photonic function.
- Heat/shielding buildings/windows.
- Smart windows, to save energy.
- Load-bearing windows that never crack or shatter.
- Translucent wood potentially transforms electronics, for example, it could be used to manufacture solar cells, especially for large surfaces.
- Provide electromagnetic interference (EMI) shielding by incorporating Magnetic nanoparticles into translucent wood.
- Translucent wood laser (entirely organic laser).

Advantages

- It is biodegradable and environmentally friendly as regular wood.
- Translucent wood has all the strength of opaque lumber.
- It does not shatter on impact and is stronger than glass.
- It is better when at light distribution and eliminates glare.
- It helps keep a building at more consistent temperature and make it easier to reach higher energy efficiency. This is because wood has low thermal conductivity.
- Another advantage of translucent wood over glass for structural applications is that the ductility and work to fracture are higher.

2.1.7 Laminated Timber (LT)

- Cross-laminated timber is a wood panel product made from gluing layers of solid-sawn lumber together. Each layer of boards is usually oriented perpendicular to adjacent layers and glued on the wide faces of each board, usually in a symmetric way so that the outer layers have the same orientation.
- An odd number of layers are most common, but there are configurations with even numbers as well (which are then arranged to give a symmetric configuration). Regular timber is an anisotropic material, meaning that the physical properties change depending on the direction at which the force is applied.
- By gluing layers of wood at right angles, the panel is able to achieve better structural rigidity in both directions. It is similar to plywood but with distinctively thicker laminations (or lamellae).
- LT is distinct from glued laminated timber, a product with all laminations orientated in the same way.
- These laminates are finding wider applications for flooring, wall paneling in door and window frames, furniture, etc. as alternate for wood and wood based products. The boards have a natural elegant pattern with beautiful colors.
- The panels are treated with anti-insects and anti–decay chemicals and bonded with urea or phenol resin adhesives depending upon end use application.

Manufacturing

- The manufacturing of LT can be split-up into nine steps: Primary lumber selection, lumber grouping, lumber planning, lumber cutting, adhesive application, panel lay-up, assembly pressing, quality control and finally marketing and shipping.
- The primary lumber selection consists of two to three parts, moisture content check, visual grading and sometimes depending on the application structural testing. Depending on the results of this selection, the timber fit for LT will be used to create either construction grade LT or appearance grade LT. Timber that cannot fit into either category may be used for different products such as plywood or glued laminated timber.

Advantages

LT has some advantages as a building material, including:

- It can be used in walls, roofs or ceilings. The thickness of the panels can easily be increased by adding more layers and the length of the panels can be increased by joining panels together.
- Eco-friendly – LT is a renewable, green and sustainable material, since it is made out of wood and does not require the burning of fossil fuels during production.
- Prefabrication – Floors or walls made from LT can be fully manufactured before reaching the job site, which decreases lead times and could potentially lower overall construction costs.
- Thermal insulation – Being made out of multiple layers of wood, the thermal insulation of LT can be high depending on the thickness of the panel.

- LT is a relatively light building material – Foundations do not need to be as large and the machinery required on-site is smaller than those needed to lift heavier buildings materials. These aspects also provide the additional capacity to erect LT buildings on sites that might otherwise be incapable of supporting heavier projects, and eases infilling projects where construction is especially tight or difficult to access due to the preexisting buildings around the site.

Disadvantages

LT also has some disadvantages, including:

- Higher production costs – Being a relatively new material, LT is not produced in many locations. Also, the production of LT panels requires a considerable amount of raw materials compared to regular stud walls.
- Limited track record – LT is a relatively new material, so it has not been used in many building projects. A considerable amount of technical research has been done on LT but it takes time to integrate new practices and results into the building industry because of the building industry's path-dependent culture which resists deviating from established practices.
- Acoustic performance – In order to achieve acceptable acoustic performance, more LT panels must be used. According to the LT handbook, two LT panels with mineral in-between achieve the international building requirement for sound insulation in walls.

2.1.8 3D Tiles

- 3D Tiles are an open specification for streaming massive heterogeneous 3D geospatial datasets.
- 3D Tiles are used to stream 3D content, including buildings, trees, point clouds, and vector data on flooring tiles, roofing tiles and wall tiles.
- These tiles give a decorative look to the building units.

2.2 ROAD MATERIALS

Needs for advanced high-performance materials.

- The needs for seeking advanced road construction materials include:
- Reduced costs – get more lane-miles constructed or rehabilitated for a given constrained budget.
- Conservation of resources – support national efforts to create sustainable solutions to minimize impact of construction on the environment.
- Reduced ecological footprint.
- Extended service life.
- Optimized use of locally available materials.
- Achieving environmental benefits – reduced carbon footprint, reduced congestion-related emissions.
- Reduced work zone–related traffic delays and safety concerns – use materials that reduce the potential for early failures.

2.2.1 Geo-synthetics

- Geo-synthetics are an established family of geomaterials used in a wide variety of civil engineering applications. Many polymers (plastics) common to everyday life are found in geo-synthetics.
- The most common are polyolefins and polyester; although rubber, fiberglass, and natural materials are sometimes used. Geo-synthetics may be used to function as a separator, filter, planar drain, reinforcement, cushion/protection, and/or as a liquid and gas barrier. The various types of geo-synthetics available, along with their specific applications.

- Geo-synthetics are synthetic products used to stabilize terrain. They are generally polymeric products used to solve civil engineering problems. This includes eight main product categories : geotextiles, geogrids, geonets, geomembranes, geofoam, geocells and geocomposites.
- The polymeric nature of the products makes them suitable for use in the ground where high levels of durability are required. They can also be used in exposed applications. Geo-synthetics are available in a wide range of forms and materials.
- These products have a wide range of applications and are currently used in many civil, geotechnical, transportation, geoenvironmental, private development applications including roads, airfields, railroads, embankments, retaining structures, reservoirs, canals, dams, erosion control, sediment control, landfill liners, land fill covers, mining, aqua culture and agriculture.

Geo-synthetics Advantages

Geo-synthetics, including geotextiles, geomembranes, geonets, geogrids, geocomposites and geo-synthetic clay liners, often used in combination with conventional materials, offer the following advantages over traditional materials :

- Space Savings - Sheet-like, geo-synthetics take-up much less space in a landfill than do comparable soil and aggregate layers.
- Material Quality Control - Soil and aggregate are generally heterogeneous materials that may vary significantly across a site or borrow area. Geo-synthetics on the other hand are relatively homogeneous because they are manufactured under tightly controlled conditions in a factory. They undergo rigorous quality control to minimize material variation.
- Construction Quality Control – Geo-synthetics are manufactured and often factory "prefabricated" into large sheets. This minimizes the required number of field connections, or seams. Both factory and field seams are made and tested by trained technicians. Conversely, soil and aggregate layers are constructed in place and are subject to variations caused by weather, handling and placement.
- Cost Savings – Geo-synthetic materials are generally less costly to purchase, transport and install than soils and aggregates.
- Technical Superiority – Geo-synthetics have been engineered for optimal performance in the desired application.
- Construction Timing – Geo-synthetics can be installed quickly, providing the flexibility to construct during short construction seasons, breaks in inclement weather, or without the need to demobilize and remobilize the earthwork contractor.
- Material Deployment - Layers of geo-synthetics are deployed sequentially, but with a minimum of stagger between layers, allowing a single crew to efficiently deploy multiple geo-synthetic layers.
- Material Availability - Numerous suppliers of most geo-synthetics and ease of shipping insure competitive pricing and ready availability of materials.
- Environmental Sensitivity – Geo-synthetic systems reduce the use of natural resources and the environmental damage associated quarrying, trucking, and other material handling activities.

Properties

1. Woven geo-textiles have high tensile strength and low strain or limited elongation under load.
2. Geo-grids have a low strain and stretch only 2 to 5% under loads. They take-up heavy loads.
3. While geotextiles, geo-grids and geo-cells are porous to allow water to filter through them. Geo-membranes have low permeability and are used to control fluid movement.
4. Geo-synthetic materials such as tubes, mats are used for erosion control. They are biodegradable and include straw and coconut malting.

Uses

- To improve level grade soil situations such as roads, valleys, laneways.
- To improve sloped-grade situations such as banks, hill sides.
- Reinforced soil: Soil walls, bridge abutments, box culverts, bridges and soil arches.
- Prevent soil movement (piping), while letting water move through the material.
- Control water pressure allowing flow in the plane of material such as foundation walls.

Following are the types of geo-synthetic materials :

1. Geo-textiles used for drainage, separation and reinforcement.
2. Geo-grids are open mesh-like materials used for stabilization and reinforcement.
3. Geo-cells are cavity like materials in a web used for stabilization.
4. Geomembranes are low permeability liner or fluid containment materials.
5. Erosion control materials that are bio degradable or non-biodegradable.

2.2.2 Noise-reducing Asphalt

- Asphalt pavement use in road design: One important way noise can be reduced is by using asphalt for pavement.
- A dense-graded hot-mix asphalt (HMA) pavement is as quiet at 100 ft. away from the road as Portland cement concrete (PCC) is at 200 ft. away. One asphalt mix, called open-graded friction courses (OGFC), is especially helpful in reducing noise because it is porous.
- It has a higher ratio of air voids, which helps water run-off but also helps absorb road noise.
- Asphalt pavement can be made even quieter and even more eco-friendly when recycled tires are mixed in–this is called rubberized asphalt.
- Noise tests on rubberized asphalt have shown a 50%-90% reduction in noise.
- Quiet asphalt pavement can help reduce highway noise by as much as 7 decibels, according to a study. Reducing noise by just 3 decibels is equivalent to doubling the distance from the source of noise to the listener, according to Asphalt Pavement literature.
- Quiet asphalt pavement options fall into four general categories. These are fine-graded surfaces, open-graded friction courses, rubberized asphalt and stone-matrix asphalt.
- Fine-graded surfaces are conventional asphalt mixes that are mostly composed of fine-grained particles; reducing the apparent road noise generated by the expelled air by altering its frequencies.
- Open-graded friction courses (OGFC) are a porous mix; with more connected voids through which air and water can escape. Rubberized asphalt contain crumbs of recycled tires that provide some flexing in the road surface as tires pass over it, allowing the air a bit more time to be forced out at lower pressure. Stone-matrix asphalt mainly uses coarse aggregate to create a stone skeleton, tied together by asphalt binders and fibers.

2.2.3 Porous Pavement

- Permeable pavement is a porous urban surface composed of open pore pavers, concrete, or asphalt with an underlying stone reservoir. Permeable pavement catches precipitation and surface run-off, storing it in the reservoir while slowly allowing it to infiltrate into the soil below or discharge via a drain tile. The most common uses of permeable pavement are parking lots, low-traffic roads, sidewalks, and driveways.
- Permeable pavement surfaces are made of either a porous material that enables storm water to flow through it or nonporous blocks spaced so that water can flow between the gaps. Storm water flows into and is stored in an underlying stone reservoir.
- Permeable pavement is commonly used on roads, paths and parking lots subject to light vehicular traffic, such as cycle-paths, service or emergency access lanes, road and airport shoulders, and residential sidewalks and driveways.

- Permeable pavement is a porous urban surface composed of open pore pavers, concrete, or asphalt with an underlying stone reservoir.
- Permeable pavement catches precipitation and surface run-off, storing it in the reservoir while slowly allowing it to infiltrate into the soil below or discharge via a drain tile.
- Permeable pavement is a porous urban surface composed of open pore pavers, concrete, or asphalt with an underlying stone reservoir.
- Permeable pavement catches precipitation and surface runoff, storing it in the reservoir while slowly allowing it to infiltrate into the soil below or discharge via a drain tile.
- Permeable paving is a method of paving vehicle and pedestrian pathways to enable infiltration of storm water run-off. Permeable pavement surfaces typically include pervious concrete, porous asphalt, paving stones and interlocking pavers.

General hydrologic benefits

- Permeable pavements help reestablish a more natural hydrologic balance and reduce run-off volume by trapping and slowly releasing precipitation into the ground instead of allowing it to flow into storm drains and out to receiving waters as effluent. This same process also reduces the peak rates of discharge by preventing large, fast pulses of precipitation through the stormwater system.
- Permeable pavement can reduce the concentration of some pollutants either physically (by trapping it in the pavement or soil), chemically (bacteria and other microbes can break down and utilize some pollutants), or biologically (plants that grow in-between some types of pavers can trap and store pollutants).
- By slowing down the process, permeable pavements can cool down the temperature of urban run-off, reducing the stress and impact on the stream or lake environment.
- By controlling the runoff at the source, such as a parking lot, permeable pavement can also reduce the need for or the required size of a regional BMP, such as a wet detention pond, which saves money and effort.

2.2.4 Plastic Roads

- Plastic roads are made entirely of plastic or of composites of plastic with other materials. Plastic roads are different from standard roads in the respect that standard roads are made from asphalt concrete, which consists of mineral aggregates and asphalt.
- Plastic-bitumen composite roads have better wear resistance than standard asphalt concrete roads. They do not absorb water; have better flexibility which results in less rutting and less need for repair. Road surfaces remain smooth, are lower maintenance, and absorb sound better.
- Plastic use in road construction is not new. It is already in use as PVC or HDPE pipe mat crossings built by cabling together PVC (polyvinyl chloride) or HDPE (high-density poly-ethylene) pipes to form plastic mats. The plastic roads include transition mats to ease the passage of tyres up to and down from the crossing. Both options help protect wetland haul roads from rutting by distributing the load across the surface. But the use of plastic-waste has been a concern for scientists and engineers for a quite long time.
- Recent studies in this direction have shown some hope in terms of using plastic-waste in road construction i.e., Plastic roads.
- Plastic roads mainly use plastic carry-bags, disposable cups and PET bottles that are collected from garbage dumps as an important ingredient of the construction material. When mixed with hot bitumen, plastics melt to form an oily coat over the aggregate and the mixture is laid on the road surface like a normal tar road.
- Plastic coating on aggregates is used for the better performance of roads. This helps to have a better binding of bitumen with plastic wasted coated aggregate due to increased bonding and increased area of contact between polymers and bitumen.

- The polymer coating also reduces the voids. This prevents the moisture absorption and oxidation of bitumen by entrapped air. This has resulted in reducing rutting, raveling and there is no pothole formation

2.2.5 Solar Roads

- The Solar Roadway produces clean, renewable energy. No pollution, no greenhouse gases, no by-products, and the Solar Road Panels are completely recyclable or reusable.
- The main cause of global warming is creation of electricity by fossil fuels which will contribute to production of greenhouse gases and effect on ozone layer.
- The Solar Roadways eliminates this (half of the cause of Global Warming) entirely. The Solar Roadway, being an "electric road", will also make all-electric vehicles more practical; recharging stations can be placed in all parking lots and rest stops.
- This will allow the all-electric vehicles to have the same driving range of current internal combustion gasoline- or diesel powered vehicles. Elimination of internal combustion engines, which would now be feasible with the Solar Roadways, would wipe out most of the rest of the causes of greenhouse gases.
- There are many other features, including wildlife preservation, the elimination of impervious surfaces, law enforcement, DUI detection, counter-terrorism etc.
- An intelligent highway infrastructure and a self-healing decentralized power grid that will eliminate our need for fossil fuels and also it will lead to less invest in antiquated technology viz asphalt and overhead power lines.
- As the day-by-day the price of petroleum products are getting huge hike and resources are very less there will be no longer feasible material such as asphalt for our road surfaces.
- When Solar Road Panels are refurnished, the solar cells will be upgraded to newest technology, which will allow keeping up with population growth and increased energy needs. Also if such technology is furnished in any of the country; the country will require approximately five billion solar road panels for covering roads, parking lots, drive ways etc. and also such technology will create millions of "Green color" jobs.
- The solar Roadways can save the wonderful countries in the world. The day by day the human beings are looking for the answers to our deteriorating highway infrastructure, our crumbling power grid, and the climate crisis. For all such questions the answer is "Solar Roadways".
- The Solar Roadways consists of structurally engineered solar panels that we drive on. Each Solar Road Panel (roughly 12′ by 12′) interlinks with neighboring panels to form the Solar Roadways system.
- The Solar Roadway replaces our crumbling petroleum-based asphalt highway infrastructure with an intelligent road that pays for itself through the generation of electricity.
- The Solar Roadway generates electrical power from the sun and becomes our nations decentralized, intelligent, self-healing power grid, replacing our current deteriorating power distribution infrastructure.
- The Solar Roadway distributes its electrical power to all businesses and homes connected to the system via their parking lots and driveways (made up of Solar Road Panels).
- In addition to electrical power, data signals (cable TV, high-speed internet, telephone etc.) also travel through the Solar Roadways, which acts as a conduit for these signals (cables). This feature eliminates the unsightly power lines, utility poles, and relay stations we see all over the countryside.
- It also eliminates power interruption caused by fallen or broken electrical lines or poles. Each Solar Road Panel uses some of its own power to light up embedded LEDs, which "paint" the road lines from beneath the road surface.
- This feature also allows messages to be spelled out on the road surface, such as "SLOW DOWN", or "ACCIDENT AHEAD". Road lines can be instantly "repainted" to direct traffic to a single lane or to detour. This eliminates the need for cones or flares. Better visibility at night with the road lines illuminated, it will be like driving on a well-lit runway.

- Solar Roadways (SR) is a modular system of specially engineered solar panels that can be walked and driven upon. Our panels contain LED lights to create lines and signage without paint. They contain heating elements to prevent snow and ice accumulation.
- The applications for solar panels are growing exponentially. Photovoltaic cells work by using semiconductors to absorb light and create a flow of electrons, which can power any number of electrical devices. Put the two together and you get a unique solution: solar panels on our highways.
- Solar Roadway panels are made of tempered (safety) glass. Glass was chosen for its hardness, strength, durability, and transmittance. SR glass is textured to create proper traction for vehicles and pedestrians.

2.2.6 Anti Icing Roads

- **Anti-icing :** The application of chemicals to roads before a snow-pavement bond occurs. Anti-icing emphasizes prevention rather than reaction.
- Anti-icing is a pro-active approach to winter road maintenance. It forms a bond-breaker between the pavement surface and the snow and ice layer which melts snow more quickly and reduces the chance that ice will form and bond to the surface. Sodium chloride (salt), magnesium chloride, calcium chloride, calcium magnesium acetate and potassium acetate are chemicals used to prevent and remove snow and ice from roadways. VDOT uses liquid magnesium chloride, calcium chloride and sodium chloride for anti-icing and pre-treatment.
- Anti-icing is a pro-active approach to winter road maintenance. It involves the application of anti-icing products (also known as freezing-point depressants) to the roadway before a winter storm.
- It forms a bond-breaker between the pavement surface and the snow and ice layer which melts snow more quickly and reduces the chance that ice will form and bond to the surface. It is similar to how cooking oil prevents food from sticking to the frying pan.
- Anti-icing reduces the amount of time required to restore the roads to a clear, dry state. NDDOT uses snow plows, tankers, automatic bridge deck sprayers, and other equipment in the anti-icing process.
- Effective anti-icing is dependent upon having the right amount of anti-icing liquid in the right place at the right time. Applying anti-icing liquids to the roadways is generally the most effective approach to prevent icy and snow-compacted roads.
- There are many factors that go into determining if anti-icing will be effective. Some of the factors are air temperature, wind speed, precipitation type and intensity, current pavement conditions, and location. NDDOT uses these factors in an advanced weather forecasting system, as well as the experience of our maintenance crews, to determine when and where to anti-ice.
- Anti-icing is a pro-active approach to winter road maintenance. It forms a bond-breaker between the pavement surface and the snow and ice layer which melts snow more quickly and reduces the chance that ice will form and bond to the surface. It reduces the amount of time required to restore the roads to a clear, dry state.
- Reduced use of abrasives on the road results in reduced environmental impacts.
- Snow and ice control cost savings results in benefits to NDDOT and the public.
- Improved winter roadway conditions results in safer driving conditions for motorists.
- Lower accident rates – Colorado experienced an average decrease of 14% in snow- and ice-related crashes during a 12-year study utilizing the anti-icing process on the interstate system.
- NDDOT applies liquid products directly to the roadway to prevent the build-up of snow and ice by lowering the freezing point of water, leaving the roads bare and wet when they would otherwise be covered by snow and ice.

- NDDOT primarily uses liquids in three ways:
  1. Anti-icing is the application of liquids to the roadway before, or at the onset to, a winter storm.
  2. De-icing is the application of liquids after snow and ice has compacted to the roadway. NDDOT's experience has been that it takes approximately three times the resources to remove compacted snow and ice when the anti-icing process is not used. The same effect is experienced when snow compacts on a driveway after driving on it several times and then trying to scrape it off.
  3. Pre-wetting is the application of liquid to a granular salt mixture to help jump start the melting process. Sand alone cannot melt snow and ice, and dry salt must change into a liquid solution to melt snow and ice. Studies have shown up to 30% of dry salt or sand can scatter or blow-off the roadway, so liquid is added to the salt mixture to prevent scattering and blowing, helping the salt mixture stay on the roadway.

2.2.7 Piezoelectric Roads

- Alternative energy will become increasingly important as fossil fuel supplies inevitably run out or environmental damage sparks consumer awareness.
- The search for a viable energy alternative will continue until these alternatives can address the dynamic demands of the electrical grid and storage limitations.
- Piezoelectric devices, used for harvesting the vibrational energy of roads and walkways due to traffic, can produce electrical energy that is predictable (based on traffic patterns), and locally storable.
- Piezoelectric devices generate electrical energy by means of a piezoelectric crystal electric floors are designed to capture the wasted energy and resources, and store or redistribute them where they are needed. Energy is generated when a person steps on tiles that feature piezoelectric attributes. This kinetic energy is converted into electricity.
- The crystal, placed about 5 centimeters below the surface of the asphalt, slightly deforms when vehicles travel across the road, thereby producing electrical current.
- The generating capacity of piezoelectric devices can be crudely over-approximated by assuming that the vibrations in the road are caused by traffic alone, and that each "vibration event" from one vehicle is independent of another (i.e. the vibrations are sufficiently dampened before the next vehicle passes).
- Under these assumptions, the total energy harvested by piezoelectric devices along a one-kilometer stretch is at most the number of cars that pass multiplied by the vibrational energy that one car transfers to the road.
- This vibrational energy can be over-approximated by the energy that each car consumes and puts to mechanical work across this stretch. In other words, the energy a car loses to vibrations in asphalt must be less than the energy a car puts to mechanical work over the one-kilometer stretch. This value can be computed by multiplying the energy consumed from gasoline by thermal efficiency.
- Having met the power metrics will make such invention a starting point to self-sustaining environment by being dependent on vehicle elements to curb down input energy and cost. This will be a revolution in the production of energy by meeting innovations and technologies to shape future energy. However, further concentration is required to quantify and result in a greater power outcome to become a reliable source of electricity.

Limitations

- Piezoelectric road integrated technology is a new energy evolution and hence piezoelectric awareness and expertise in this sector are very limited. This was also observed in primary data collected using questionnaires and interview to have obtain invalid and bias information.
- The power metrics provided by vendors does not provide additional technical details. The vendors were compared using vibration-based factor only. Other technical factors were not analysed in terms of piezoelectric geometry, structure, and thickness. This is common having not to reveal a full technical report of this technology to preserve innovation, invention rights and securitise effective business strategies.
- Implementing piezoelectric road technology on field requires a standard specification in the execution process which has not been established. This is important as to use appropriate management and method to prevent manipulating with road infrastructure and minimize traffic congestion.

2.3 CONCRETE MATERIALS

- The aim of development of new concrete technologies was the requirement for high durability of concrete structures.

2.3.1 Portland Pozzolana Cement

- Portland Pozzolana cement is integrated cement which is formed by synthesising OPC cement with pozzolanic materials in a certain proportion. It is commonly known as PPC cement.
- Pozzolana is a volcanic powder found in Italy near Vesuvius. A pozzolanic material can be a natural or artificial which contains silica and aluminous in a reactive form. This materials usually doesn't possess any cementitious properties, but when it is mixed with water or moisture or lime to undergo reaction with calcium hydroxide to form compounds possessing cement properties.

Types of pozzolana materials:

1. **Artificial pozzolana**

- Fly ash, silica fume, rice husk, blast furnace slag.

2. **Natural pozzolana**

- Burnt clay, pumicite, diatomaceous Earth.

Manufacture of Portland pozzolana cement

- The primary raw materials used for this cement manufacture are limestone ($CaCO_2$) and clay (SiO_2, Al_2O_3, Fe_2O_3). Rocks are loaded into trucks and transported to the crushers, where the crushed into fine particles.
- Fine particles of clay and limestone are fed into the air-swept ball mills in desired proportions as per requirement and mixed very well before it is sent to silos for storing.
- This mixture is then pre-heated upto 800-1000c where calcinations of $CACO_3$ to CaO takes place.
- The pre heated mixture is then sent into kiln where the mixture is heated to 145°C in rotary Kiln. The modules formed from the burning process called clinker. The clinker is cooled by a rotary cooler.
- This clinker is now mixed with gypsum and pozzolana materials in the required proportion and thus the Portland Pozzolana Cement is obtained.

Properties of Portland Pozzolana Cement

- Initial setting time = 30 min (minimum)
- Final setting time = 600 min (maximum).
- At 3 days 13MPa (minimum)
- At 7 days 22 MPa (minimum)

- At 28 days 33 MPa (minimum)
- Drying shrinkage should not be more than 0.15%
- Fineness should not be less than 300 m^2/kg
- Initial strength of PPC is less but final strength is equal to the 28 days strength of OPC
- PPC has lower rate of development of strength than OPC

Uses of Portland Pozzolana Cement

- Used in hydraulic structures, marine structures, construction near the sea shore, dam construction etc.
- Used in pre-stressed and post-tensioned concrete members.
- Used in masonry mortars and plastering.
- As it gives better surface finish, it is used in decorative and art structures.
- Used in manufacture of precast sewage pipes.
- Used under harsh concreting conditions.

Advantages of Portland Pozzolana Cement

- It is an eco-friendly cement as the material used in the manufacture are made of natural recycled waste.
- It is very fine cement hence very good when used for plastering works.
- Pozzolano consists of silica material which makes it cheap and hence reduces the cost of the cement making it economical to use.
- Pozzolana cement has very good resistance against sulphate attack hence is used in hydraulic structures, marine structures, construction near the sea shore, dam construction etc.
- PPC used in pre-stressed and post-tensioned concrete members.
- It reduces the carbon monoxide emission from the concrete making it environmental friendly.
- As the pozzolano materials are very fine, it can fill gaps between the reinforcement and aggregate, thus reducing the shrinkage, honeycomb formation and bleeding can be reduced, which in turn increases the strength and durability of concrete.

Disadvantages of Portland Pozzolana Cement

- The initial strength obtained is less, which effect the de-shuttering of supports early.
- As it contains more fine material, handing of concrete is difficult.
- When compared to the OPC setting time is less for PPC
- Reduction in alkanity reduces the resistance to corrosion of steel reinforcement
- As the strength of this concrete gains slowly, curing process is very important. Any error in this could cause durability problems.

2.3.2 Portland Slag Cement

- This cement is produced by mixing Portland cement clinker, gypsum and granulated Blast Furnace slag in suitable proportions and grinding the mixture to get a thorough and intimate mixture between the constituents.
- According to the control over physical and chemical specification, slag is mixed up to 25-65%. Slag is a byproduct in the manufacture of pig iron.
- Chemically, slag is a mixture of lime, silica and alumina, that is, the same oxides that make-up Portland cement.
- Using slag cement to replace a portion of Portland cement in a concrete mixture is an efficient way to make concrete more consistent. Among the measurable improvements are higher workability, higher finishability, and lower permeability, improved resistance to aggressive chemicals, more consistent plastic and hardened properties, and flexural strengths. Concrete with slag is easier to work and less susceptible to freeze thaw cycles.

Advantages to use portland slag cement

- **Prevention against Chemical Hazards**: Use of Slag Cement significantly reduces the risk of damages caused by alkali–silica reaction (ASR), provides higher resistance to chloride ingress — reducing the risk of reinforcement corrosion — and provides higher resistance to attacks by soils and water containing excess amount of sulfates or alkali metals, alumina and iron, as well as to acidic waters.
- **Strength Development**: Concrete containing Slag cement has a higher ultimate strength than concrete made with Portland cement. It has a higher proportion of the strength-enhancing calcium silicate hydrates (CSH) than concrete made with Portland cement only, and a reduced content of free lime, which does not contribute to concrete strength. Concrete made with Slag cement continues to gain strength over time, and has been shown to double its 28-day strength over periods of 10 to 12 years. Concrete containing ground granulated slag develops strength over a longer period, leading to reduced permeability and better durability.
- **Lower Permeability**: Being Slag cement very fine, the concrete made by using slag cement has better resistuvity against permeability.
- **Durability**
  Greater resistance to chlorides and sulfates can help to reduce corrosion of steel reinforcement and deterioration of the structure. In most developed countries, is now specified in structural concrete for bridge piers and abutments for protection against chloride attack. The use of Slag in such instances will increase the life of the structure by up to 50% had only Portland cement been used, and precludes the need for more expensive stainless steel reinforcing.
- Slag is also routinely used to limit the temperature rise in large concrete pours. The more gradual hydration of Slag cement generates both lower peak and less total overall heat than Portland cement. This reduces thermal gradients in the concrete, which prevents the occurrence of micro cracking which can weaken the concrete and reduce its durability.
- **Appearance**: In contrast to the stony grey of concrete made with Portland cement, the near-white color of Slag cement permits architects to achieve a lighter colour for exposed fair-faced concrete finishes, at no extra cost. To achieve a lighter color finish, Slag is usually specified. Slag cement also produces a smoother, more defect free surface, due to the fineness of the Slag particles. Dirt does not adhere to Slag Cement concrete as easily as concrete made with Portland cement, reducing maintenance costs. Slag cement prevents the occurrence of efflorescence, the staining of concrete surfaces by calcium carbonate deposits. Due to its much lower lime content and lower permeability, Slag is effective in preventing efflorescence when used at appropriate levels.
- **Eco-friendly**: From an environmental viewpoint, the use of slag cement in concrete serves to make concrete "greener." Not only is slag cement can be considered a recycled material, but it can also significantly reduce energy consumption and greenhouse gasses emitted in the production of concrete raw materials.
- This previously unwanted recycled product is used in the manufacture of high performance concretes.

2.3.3 New Admixtures

- The biggest change in concrete in India occurred with the introduction of a new concrete technology that greatly expanded the plastic and hardened properties of concrete and, in the case of Masterglanium and Polycarboxylic Ether created a new concrete terminology.

2.3.3.1 Masterglanium

- The Masterglanium product offering comprises new generation high-range water-reducing admixtures that are specially formulated for applications where slump retention, high/early strengths and durability are required.
- Concrete mixtures containing these premier products can be optimized for delivery in remote locations and for use in hot and cold climates.

- The unrivalled technology and versatility of the Masterglanium product portfolio excels in high performance areas such as high strength concrete, pumpability, self-consolidating concrete, low permeability, low water absorption concrete and low shrinkage applications. Its technology provides excellent concrete performance in both the fresh and hardened states with appealing surface appearances.
- Based on the latest generation of polymer technology, the Masterglanium admixture portfolio is made up of many customized and tailored molecules. This wide-range of solutions provides an unmatched space for producers to meet performance and placement challenges.
- The unrivalled technology and versatility of the Masterglanium product portfolio excels in high performance areas such as high strength concrete, pumpability, self-consolidating concrete, low permeability, low water absorption concrete and low shrinkage applications. Its technology provides excellent concrete performance in both the fresh and hardened states with appealing surface appearances.
- Based on the latest generation of polymer technology, the Masterglanium admixture portfolio is made-up of many customized and tailored molecules. This wide-range of solutions provides an unmatched space for producers to meet performance and placement challenges.
- Super plasticizer by trade name Masterglanium was used.
- Masterglanium is an innovative versatile admixture based on third generation polycarboxylic ether (PCE) polymers.
- MasterGlanium is a high range water reducing admixtures. It may be used in combination with MasterMATRIX admixtures for producing Rheodynamic concrete, capable of self-compaction, even in the presence of dense reinforcement.

2.3.3.2 Polycarboxylic Ether

- Now, the interesting part about this polycarboxylic ether is that you have a side chain that is extremely bulky. So that means you have a main chain which is long. In all the other polymers, you have a long chain structure for the polymer.
- But in the polycarboxylic ether and the third generation admixtures, you have very bulky side chains coming from the long chain. So bulky side chains are coming out from the long chain polymers and this bulky side chain is what gives the new generation admixtures are slightly different mechanism of action.
- The polycarboxylic ether acid has a very interesting sort of a structure here with very bulky side chain. So again you have surface adsorption; the long chain polymer molecules attach themselves to the surface of the cement, thus orienting their bulky side chain outward.
- So what is happening now is; after you have the adsorption taking place, the structure of the polymer itself prevents the cement grains from getting together. In other words, flocculation is prevented by the bulkiness of the side chain of the chemical admixture and this process has got steric hindrance.
- So, in the older generation admixtures, the repulsion was being caused because of charge. In this case the repulsion is being caused because of steric hindrance; that is the difference between old and new generation. Now, problem with the old generation is the charge will diminish after some time. So the effectiveness of the older generation admixtures in maintaining the slump would be low.
- Whereas the new generation admixtures tend to maintain slump for a longer time because they dont depend of surface charge, they depend on the availability of bulky side chains in the material. So, the hydrates basically, will start filling-up the structure and then your charge will go down in case of old admixtures.
- When water starts reacting with the cement, your charge will diminish but in case of polycarboxylic ether, the effectiveness of the chemical will last for longer time because you are not relying on the charge.

- Other kinds of chemicals that will have this type of an action are cross linked acrylic polymers or carboxylic acid with acrylic ester. These chemicals can actually provide you up to 40% water reduction and fairly good slump retention.
- You all know about world record for concrete pumping for self-compacting concrete is with Burj Khalifa; 600 metres vertically upwards and we can imagine that the time taken for pumping the concrete up and then applying the concrete on site in a self-compacting form would have been tremendously large and the slump had to be maintained for a long period of time and that was done effectively with the use of polycarboxylic ether based super plasticizer.
- So, polycarboxylic ether admixture or PC admixtures act by steric hindrance because of which their effective at low water cement ratios.
- Especially when you want to produce very high grades of concrete or ultra-high strength concrete, you have to rely on polycarboxylic ether admixture and generally they are not known to have compatibility problem.
- With sulphonated naphthalene formaldehyde, if you are providing a very high molecular weight chemical, you may still get some steric hindrance but its secondary. The primary aspect there is the electrostatic repulsion and not the steric hindrance.

2.3.4 Self Healing Concrete

- This common problem of cracking in building has many remedies before and after the crack. One of the remedial process is Self-Healing Concrete.
- The process of self-healing of cracks or self-filling up of cracks by the help of bacterial reaction in the concrete after hardening is known as Self-Healing Concrete.
- It can be observed that small cracks that occur in a structure of width in the range of 0.05 to 0.1 mm gets completely sealed in repetitive dry and wet cycles. The mechanism of this autogenously healing is, the width of range 0.05-0.1 mm act as capillary and the water particles seep through the cracks. These water particles hydrate the non or partial reacted cement and the cement expands, which in turn fills the crack.
- But when the cracks are of greater width, need of other remedial work is required. One possible technique is currently being investigated and developed was based on application of mineral producing bacteria in concrete.
- The bacteria used for self-healing of cracks are acid producing bacteria. These types of bacteria can be in dormant cell and be viable for over 200 years under dry conditions. These bacteria acts as a catalyst in the cracks healing process.

Various Types of Bacteria Used in Concrete

There are various types of bacteria were used in bacterial concrete construction are:

- Bacillus pasteurizing.
- Bacillus sphaericus.
- Escherichia coli.
- Bacillus subtilis.
- Bacillus cohnii.
- Bacillus balodurans.
- Bacillus pseudofirmus.

Mechanism of Bacterial Concrete

- Self-healing concrete is a result of biological reaction of non-reacted limestone and a calcium based nutrient with the help of bacteria to heal the cracks appeared on the building.
- Special type of bacteria's known as Bacillus are used along with calcium nutrient known as Calcium Lactate. While preparation of concrete, this products are added in the wet concrete when the mixing is done. This bacteria's can be in dormant stage for around 200 years.

- When the cracks appear in the concrete, the water seeps in the cracks. The spores of the bacteria germinate and starts feeding on the calcium lactate consuming oxygen. The soluble calcium lactate is converted to insoluble limestone. The insoluble limestone starts to harden. Thus filling the crack, automatically without any external aide.
- The other advantage of this process is, as the oxygen is consumed by the bacteria to convert calcium into limestone, it helps in the prevention of corrosion of steel due to cracks. This improves the durability of steel reinforced concrete construction.

Preparation of Bacterial Concrete

Bacterial concrete can be prepared in two ways,

- ➢ By direct application.
- ➢ By encapsulation in lightweight concrete.

- In the direct application method, bacterial spores and calcium lactate is added into concrete directly when mixing of concrete is done. The use of this bacteria and calcium lactate doesn't change the normal properties of concrete. When cracks are occurred in the structure due to obvious reasons.
- The bacteria are exposed to climatic changes. When water comes in contact with this bacteria, they germinate and feed on calcium lactate and produces limestone. Thus sealing the cracks.
- By encapsulation method the bacteria and its food i.e. calcium lactate, are placed inside treated clay pellets and concrete is prepared. About 6% of the clay pellets are added for making bacterial concrete.
- When concrete structures are made with bacterial concrete, when the crack occurs in the structure and clay pellets are broken and the bacteria germinate and eat down the calcium lactate and produce limestone, which hardens and thus sealing the crack. Minor cracks about 0.5 mm width can be treated by using bacterial concrete.

Advantages of Self-healing Concrete

- Self-repairing of cracks without any external aide.
- Significant increase in compressive strength and flexural strength when compared to normal concrete.
- Resistance towards freeze-thaw attacks.
- Reduction in permeability of concrete.
- Reduces the corrosion of steel due to the cracks formation and improves the durability of steel reinforced concrete.
- Bacillus bacteria are harmless to human life and hence it can be used effectively.

Disadvantages of Self-healing Concrete

- Cost of bacterial concrete is double than conventional concrete.
- Growth of bacteria is not good in any atmosphere and media.
- The clay pellets holding the self-healing agent comprise 20% of the volume of the concrete. This may become a shear zone or fault zone in the concrete.
- Design of mix concrete with bacteria here is not available any IS code or other code.
- Investigation of calcite precipitate is costly.

2.3.5 Fibre-Reinforced Concrete

- Fiber Reinforced Concrete can be defined as a composite material consisting of mixtures of cement, mortar or concrete and discontinuous, discrete, uniformly dispersed suitable fibers. Fiber reinforced concrete is of different types and properties with many advantages. Continuous meshes, woven fabrics and long wires or rods are not considered to be discrete fibers.

- Fiber is a small piece of reinforcing material possessing certain characteristics properties. They can be circular or flat. The fiber is often described by a convenient parameter called "aspect ratio". The aspect ratio of the fiber is the ratio of its length to its diameter. Typical aspect ratio ranges from 30 to 150.
- Fiber reinforced concrete (FRC) is concrete containing fibrous material which increases its structural integrity. It contains short discrete fibers that are uniformly distributed and randomly oriented. Fibers include steel fibers, glass fibers, synthetic fibers and natural fibers. Within these different fibers that character of fiber reinforced concrete changes with varying concretes, fiber materials, geometries, distribution, orientation and densities.
- Fibre-reinforcement is mainly used in shotcrete, but can also be used in normal concrete. Fibre-reinforced normal concrete are mostly used for on-ground floors and pavements, but can be considered for a wide range of construction parts (beams, pliers, foundations etc) either alone or with hand-tied rebars
- Concrete reinforced with fibres (which are usually steel, glass or "plastic" fibres) is less expensive than hand-tied rebar, while still increasing the tensile strength many times. Shape, dimension and length of fibre is important. A thin and short fibre, for example short hair-shaped glass fibre, will only be effective the first hours after pouring the concrete (reduces cracking while the concrete is stiffening) but will not increase the concrete tensile strength.

Effect of Fibers in Concrete

- Fibres are usually used in concrete to control plastic shrinkage cracking and drying shrinkage cracking. They also lower the permeability of concrete and thus reduce bleeding of water. Some types of fibres produce greater impact, abrasion and shatter resistance in concrete. Generally fibres do not increase the flexural strength of concrete, so it can not replace moment resisting or structural steel reinforcement. Some fibres reduce the strength of concrete.
- The amount of fibres added to a concrete mix is measured as a percentage of the total volume of the composite (concrete and fibres) termed volume fraction (Vf). Vf typically ranges from 0.1 to 3%. Aspect ratio (l/d) is calculated by dividing fibre length (l) by its diameter (d). Fibres with a non-circular cross section use an equivalent diameter for the calculation of aspect ratio.
- If the modulus of elasticity of the fibre is higher than the matrix (concrete or mortar binder), they help to carry the load by increasing the tensile strength of the material. Increase in the aspect ratio of the fibre usually segments the flexural strength and toughness of the matrix. However, fibres which are too long tend to "ball" in the mix and create workability problems.
- Some recent research indicated that using fibres in concrete has limited effect on the impact resistance of concrete materials. This finding is very important since traditionally people think the ductility increases when concrete reinforced with fibres. The results also pointed out that the micro fibres is better in impact resistance compared with the longer fibres.

Necessity of Fiber Reinforced Concrete

- It increases the tensile strength of the concrete.
- It reduce the air voids and water voids the inherent porosity of gel.
- It increases the durability of the concrete.
- Fibres such as graphite and glass have excellent resistance to creep, while the same is not true for most resins. Therefore, the orientation and volume of fibres have a significant influence on the creep performance of rebars/tendons.
- Reinforced concrete itself is a composite material, where the reinforcement acts as the strengthening fibre and the concrete as the matrix. It is therefore imperative that the behaviour under thermal stresses for the two materials be similar so that the differential deformations of concrete and the reinforcement are minimized.
- It has been recognized that the addition of small, closely spaced and uniformly dispersed fibers to concrete would act as crack arrester and would substantially improve its static and dynamic properties.

Different Types of Fiber Reinforced Concrete

Following are the different type of fibers generally used in the construction industries :

- Steel Fiber Reinforced Concrete.
- Polypropylene Fiber Reinforced (PFR) cement mortar and concrete.
- GFRC Glass Fiber Reinforced Concrete.
- Asbestos Fibers.
- Carbon Fibers.
- Organic Fibers.

1. Steel Fiber Reinforced Concrete

- A no of steel fiber types are available as reinforcement. Round steel fiber the commonly used type are produced by cutting round wire in to short length. The typical diameter lies in the range of 0.25 to 0.75 mm. Steel fibers having a rectangular c/s are produced by silting the sheets about 0.25 mm thick.
- Fiber made from mild steel drawn wire. Conforming to IS : 280-1976 with the diameter of wire varying from 0.3 to 0.5 mm have been practically used in India.
- Round steel fibers are produced by cutting or chopping the wire, flat sheet fibers having a typical c/s ranging from 0.15 to 0.41 mm in thickness and 0.25 to 0.90 mm in width are produced by silting flat sheets.
- Deformed fiber, which are loosely bounded with water-soluble glue in the form of a bundle are also available. Since individual fibers tend to cluster together, their uniform distribution in the matrix is often difficult. This may be avoided by adding fibers bundles, which separate during the mixing process.

2. Polypropylene Fiber Reinforced (PFR) cement mortar and concrete

- Polypropylene is one of the cheapest and abundantly available polymers polypropylene fibers are resistant to most chemical and it would be cementitious matrix which would deteriorate first under aggressive chemical attack. Its melting point is high (about 165 degrees centigrade). So that a working temp. As (100 degree centigrade) may be sustained for short periods without detriment to fiber properties.
- Polypropylene fibers being hydrophobic can be easily mixed as they do not need lengthy contact during mixing and only need to be evenly distressed in the mix.
- Polypropylene short fibers in small volume fractions between 0.5 to 15 commercially used in concrete.

3. GFRC – Glass Fiber Reinforced Concrete

- Glass fiber is made-up from 200-400 individual filaments which are lightly bonded to make-up a stand. These stands can be chopped into various lengths, or combined to make cloth mat or tape. Using the conventional mixing techniques for normal concrete it is not possible to mix more than about 2% (by volume) of fibers of a length of 25 mm.
- The major appliance of glass fiber has been in reinforcing the cement or mortar matrices used in the production of thin-sheet products. The commonly used verities of glass fibers are e-glass used. In the reinforced of plastics and AR glass E-glass has inadequate resistance to alkalis present in Portland cement where AR-glass has improved alkali resistant characteristics. Sometimes polymers are also added in the mixes to improve some physical properties such as moisture movement.

4. Asbestos Fibers

- The naturally available inexpensive mineral fiber, asbestos, has been successfully combined with Portland cement paste to form a widely used product called asbestos cement. Asbestos fibers here thermal mechanical and chemical resistance making them suitable for sheet product pipes, tiles and corrugated roofing elements. Asbestos cement board is approximately two or four times that of unreinforced matrix. However, due to relatively short length (10 mm) the fiber have low impact strength.

5. Carbon Fibers

- Carbon fibers from the most recent and probability the most spectacular addition to the range of fiber available for commercial use. Carbon fiber comes under the very high modulus of elasticity and flexural strength. These are expansive. Their strength and stiffness characteristics have been found to be superior even to those of steel. But they are more vulnerable to damage than even glass fiber, and hence are generally treated with resign coating.

6. Organic Fibers

- Organic fiber such as polypropylene or natural fiber may be chemically more inert than either steel or glass fibers. They are also cheaper, especially if natural. A large volume of vegetable fiber may be used to obtain a multiple cracking composite. The problem of mixing and uniform dispersion may be solved by adding a superplasticizer.

2.3.6 High Strength Concrete

- High-strength concrete is typically used in the erection of high-rise structures. It has been used in components such as columns (especially on lower floors where the loads will be greatest), shear walls, and foundations. High strengths are also occasionally used in bridge applications as well.
- The primary difference between high-strength concrete and normal-strength concrete relates to the compressive strength that refers to the maximum resistance of a concrete sample to applied pressure.
- Likewise, there is not a precise point of separation between high-strength concrete and ultra-high performance concrete, which has greater compressive strength than high-strength concrete and other superior properties.
- Manufacture of high-strength concrete involves making optimal use of the basic ingredients that constitute normal-strength concrete. Producers of high-strength concrete know what factors affect compressive strength and know how to manipulate those factors to achieve the required strength. In addition to selecting a high-quality portland cement, producers optimize aggregates, then optimize the combination of materials by varying the proportions of cement, water, aggregates, and admixtures.
- When selecting aggregates for high-strength concrete, producers consider the strength of the aggregate, the optimum size of the aggregate, the bond between the cement paste and the aggregate, and the surface characteristics of the aggregate. Any of these properties could limit the ultimate strength of high-strength concrete.
- Pozzolans, such as fly ash and silica fume, are the most commonly used mineral admixtures in high-strength concrete. These materials impart additional strength to the concrete by reacting with portland cement hydration products to create additional C-S-H gel, the part of the paste responsible for concrete strength.
- It would be difficult to produce high-strength concrete mixtures without using chemical admixtures. A common practice is to use a superplasticizer in combination with a water-reducing retarder. The superplasticizer gives the concrete adequate workability at low water-cement ratios, leading to concrete with greater strength. The water-reducing retarder slows the hydration of the cement and allows workers more time to place the concrete.
- High-strength concrete is specified where reduced weight is important or where architectural considerations call for small support elements. By carrying loads more efficiently than normal-strength concrete, high-strength concrete also reduces the total amount of material placed and lowers the overall cost of the structure.
- High strength of concrete is achieved by reducing porosity, in-homogeneity, and micro-cracks in the hydrated cement paste and the transition zone. Consequently, there is a reduction of the thickness of the interfacial transition zone in high-strength concrete.
- The densification of the interfacial transition zone allows for efficient load transfer between the cement mortar and the coarse aggregate, contributing to the strength of the concrete. For very high-strength concrete where the matrix is extremely dense, a weak aggregate may become the weak link in concrete strength.

Differences between Normal Strength Concrete and High Strength Concrete

- Micro-cracks are developed in the normal strength concrete when its compressive strength reaches 40% of the strength. The cracks interconnect when the stress reaches 80-90% of the strength.
- For High Strength Concrete, Iravani and MacGregor reported linearity of the stress-strain diagram at 65 to 70, 75 to 80 and above 85% of the peak load for concrete with compressive strengths of 65, 95, and 105 MPa.
- The fracture surface in NSC is rough. The fracture develops along the transition zone between the matrix and aggregates. Fewer aggregate particles are broken. The fracture surface in HSC is smooth. The cracks move without discontinuities between the matrix and aggregates.
- High strength concrete and high-performance concrete are not synonymous because strength and performance of concrete are different properties of concrete.

2.3.7 High Performance Concrete

- High-performance means that the concrete has one or more of the following properties: low shrinkage, low permeability, a high modulus of elasticity, or high strength. High-strength concrete is typically recognized as concrete with a 28-day cylinder compressive strength greater than 6000 psi or 42 Mpa.
- The composition of high performance concrete (HPC) are almost same as those of Conventional Cement Concretes (CCC).
- But, because of lower Water Cement Ratio, presence of Pozzolans and chemical admixtures etc., the HPCs usually have many features which distinguish them from CCCs.
- From practical considerations, in concrete constructions, apart from the final strength, the rate of development of strength is also very important.
- The High performance concrete usually contains both pozzolanic and chemical admixtures. Hence, the rate of hydration of cement and the rate of strength development in HPC is quite different from that of conventional cement concrete (CCC).
- For high strength concretes, however, all the components of the concrete mixture are pushed to their limits. Therefore, it is necessary to pay careful attention to all aspects of concrete production, i.e., selection of materials, mix design, handling and placing.

The proportioning (mix design) of High Performance concrete consists of three interrelated steps:

1. Selection of suitable ingredients – cement, supplementary cementing materials (SCM), aggregates, water and chemical admixtures.
2. Determination of relative quantities of these materials in order to produce, as economically as possible, a concrete that has the rheological properties, strength and durability.
3. Careful quality control of every phase of the concrete making process.

Composition of HPC

The most common composition of high performance concrete as supplementing cementitious materials or mineral admixtures are:

1. Silica Fume.
2. Fly Ash.
3. GGBFS (Ground granulated blast furnace slag).

1. Silica Fume in HPC

- Silica fume is a waste by-product of the production of silicon and silicon alloys. Silica fume is available in different forms, of which the most commonly used now is in a densified form. In developed countries it is already available readily blended with cement.
- It is possible to make high strength concrete without silica fume, at compressive strength of upto 98 Mpa. Beyond that strength level however, silica fume becomes essential. With silica fume it is easier to make HPC for strengths between 63-98 Mpa.

2. Fly Ash in HPC

- Fly Ash of course, been used extensively in concrete for many years.
- Fly ash is, unfortunately, much more variable than silica fumes in both their physical and chemical characteristics. Most fly ashes will result in strengths of not more than 70 Mpa. Therefore for higher strengths, silica fume must be used in conjunction with fly ash.
- For high strength concrete, fly ash is used at dosage rates of about 15 % of cement content.

3. Ground granulated blast furnace slag (GGBFS) in HPC:

- Slags are suitable for use in high strength concrete at dosage rates between 15-30 %. However, for very high strengths, in excess of 98Mpa, it is necessary to use the slag in **conjunction with silica fumes.**

Key Features of High Performance Concrete (HPC)

- Compressive strength > 80 MPa, even upto 800 MPa.
- Water-binder ratio =0.25-0.35, therefore very little free water.
- Reduced flocculation of cement grains.
- Wide range of grain sizes.
- Densified cement paste.
- No bleeding–homogeneous mix.
- Less capillary porosity.
- Discontinuous pores.
- Stronger transition zone at the interface between cement paste and aggregate.
- Low free lime content.
- Endogenous shrinkage.
- Powerful confinement of aggregates.
- Little micro-cracking until about 65-70% of fck.
- Smooth fracture surface.

2.3.8 Nano Concrete

- Nano concrete is defined as a concrete made by filling the pores in traditional concrete using nano particles of size < 500 nano meters. Then the mechanical properties of concrete such as compressive strength, tensile strength and flexural strength of the respective specimens were tested after 7 days and 28 days curing.
- Nano technology in civil field has proved to make the construction faster, cheaper, safer and more varied. Automation of Nano technology construction can allow for the creation of structures from advanced home to skyscrapers much more quickly and efficiently. Amorphous nano-silica (nS) particles were incorporated in cement concrete, and their effect on the fresh state behaviour was analyzed. Nano concrete is defined as a concrete made by filling the pores in traditional concrete using nano particles of size <500nano meters.
- Nano particles of silica turn into nano particles of cement (nano cement) in the chemical reactions that take place in the concoction of the concrete.
- When concrete is reduced to Nano level its properties are strongly influenced so that it increases their strength and durability.
- Addition of nano silica to the concrete leads to improve the material passing which results in the densifying of micro and Nano structures. In this project, Nano-silica has been replaced in various proportions such as 5%, 10%, 15% and 20% to the weight of cement.
- Then the mechanical properties of concrete such as compressive strength, tensile strength and flexural strength of the respective specimens were tested after 7 days and 28 days curing. Results have been obtained and compared with conventional concrete mix. Nano concrete was concluded to have a higher strength than the ordinary concrete.

- Concrete, the most ubiquitous material in the world, is a nanostructured, multi-phase, composite material that ages over time. It is composed of an amorphous phase, nanometer to micrometer size crystals, and bound water.
- The amorphous phase, calcium-silicate-hydrate (C-S-H) is the "glue" that holds concrete together and is itself a nanomaterial. Viewed from the bottom-up, concrete at the nanoscale is a composite of molecular assemblages, surfaces (aggregates, fibres), and chemical bonds that interact through local chemical reactions, intermolecular forces, and intraphase diffusion.
- Properties characterizing this scale are molecular structure; surface functional groups; and bond length, strength (energy), and density. The structure of the amorphous and crystalline phases and of the interphase boundaries originates from this scale.
- The properties and processes at the nanoscale define the interactions that occur between particles and phases at the microscale and the effects of working loads and the surrounding environment at the macroscale. Processes occurring at the nanoscale ultimately affect the engineering properties and performance of the bulk material.
- There are two main avenues of applications of nanotechnology in concrete research; the nanoscience and nano-engineering. Nanoscience deals with the measurement and characterization of the nano and microscale structure of cementbased materials to better understand how this structure affects macroscale properties and performance through the use of advanced characterization techniques and atomistic or molecular level modelling.
- Nano-engineering encompasses the techniques of manipulation of the structure at the nanometer scale to develop a new generation of tailored, multifunctional, cementitious composites with superior mechanical performance and durability potentially having a range of novel properties such as: low electrical resistivity, self-sensing capabilities, self-cleaning, self-healing, high ductility, and selfcontrol of cracks.
- Concrete can be nano-engineered by the incorporation of nanosized building blocks or objects (for example, nanoparticles and nanotubes) to control material behaviour and add novel properties, or by the grafting of molecules onto cement particles, cement phases, aggregates, and additives (including nanosized additives) to provide surface functionality, which can be adjusted to promote specific interfacial interactions.

2.3.9 Light Transmitting Concrete

- Light transmitting concrete is achieved by replacing aggregates with transparent alternate materials. The bonding material in Light transmitting concrete may be able to transmit light by using clear resins the concrete mix. Use of optical fibers and fine concrete also used as Light transmitting concrete.
- Light transmitting concrete is produced by mixing 4% to 5% (by volume) optical fibers in the concrete mixture. This concrete has less weight compared to original concrete.

Materials for Light transmitting concrete

- Light transmitting concrete is manufactured by using combination of fiber optics and fine concrete. These fibers blend into the concrete like any other aggregates. These optical fibers can transmit light from natural and artificial sources into spaces enclosed by the translucent concrete panels. The main reason for using optical fiber in concrete is that it can transmit light even an incident angle greater than 600.
- Optical fiber consists of three layers called as core, cladding and buffer coating or jacket. The light is transmitted through the core of the optical fiber.
- Light transmitting concrete is manufactured using fine materials only. It does not contain coarse aggregates. This concrete can have the compressive strength of that of high strength concrete around 70 MPa (10,000 psi).

Materials used:

- **Cement:** As the optical fiber is only responsible for transmission of light, there is no special cement required. So, ordinary Portland cement is used for Light transmitting concrete.

- **Sand:** Since the Light transmitting concrete is manufactured only using fine materials, the size of sand should pass through 1.18 mm sieve. The sand should be free from any impurities such as vegetation, large stones etc.
- **Water:** Water to be used for Light transmitting concrete should be of drinking water quality, free from any impurities.
- **Optical fibers:** Optical fibers in the range of 4 to 5% by volume is used for Light transmitting concrete. Thickness of the optical fibers can be varied between 2 μm and 2 mm to suit the particular requirements of light transmission.

Advantages of Light transmitting concrete:

- The main advantage of Light transmitting concrete is that it can transmit light. There, it can be used to make green buildings. Since it can transmit light from natural as well as artificial sources, the building can have fewer lights to meet its demand for lighting. Thus saving huge energy cost.
- Light transmitting concrete uses sunlight as source of light instead of electrical energy and reduces power consumption. This concrete can also be used cold countries to transmit heat with sunlight.

2.4 SUSTAINABLE MATERIALS

- Sustainable material is considered smart only when it contributes something to upgrade the quality of building.
- With all those advancements in construction techniques and also with the demand of end users for the smart buildings we as constructors and designers are ought to introduce something new and smart to fulfill their demands and needs.

2.4.1 Ground Granulated Glass Blast-furnace Slag (GGBS) Concrete

- The ground granulated blast-furnace slag (GGBS) is a by-product of iron manufacturing which when added to concrete improves its properties such as workability, strength and durability.
- This material is obtained by the heating of iron ore, limestone and coke at a temperature about 1500 degree Celsius. The process is carried out in a blast furnace.
- The formation of GGBS is not direct. The by-product of iron manufacturing is a molten slag and molten iron. The molten slag consists of alumina and silica, also with the certain amount of oxides.
- This slag is later granulated by cooling it. For this, it is allowed to pass through a high-pressure water get. This result in quenching of the particles which results in granules of size lesser than 5mm in diameter.
- The main constituents of blast furnace slag are CaO, SiO_2, Al_2O_3 and MgO. These are the minerals that are found in most of the cementitious substances.
- The particles are further dried and ground in a rotating ball mill to form a fine powder, known as ground granulated blast-furnace slag cement. Now different methods can be employed to perform the main process called as the quenching.
- Based on what method is employed, it can be known as palletized slag, foamed or expanded slag, GGBS or air-cooled blast-furnace slag (ACBS).
- It is found that working with GGBS is easy as it has greater mobility characteristics. This is due to its fineness and the particle shape of the GGBFS particles. These also possess a lower relative density.
- The GGBS particles have a very glassy texture that makes them increase the workability. This can help in reduction of water as well as Superplasticizers to get adequate workability in common situations.
- They also have fewer chances to get segregated during handling as well as pumping of the material. Pumping is facilitated by the lower relative density and flowing ability of the mix, that is owned by GGBS.

- The usage of GGBS in concrete provides a certain unique feature that differs from the ordinary Portland cement concrete. The concrete made from GGBFS sets slowly when compared to ordinary Portland cement mix.
- More the GGBS amount, more will be the time taken for its setting. But the strength is gained with time. This slow setting would help in the formation of cold joints. But the situations where faster setting time is required, cannot go for this replacement.
- The GGBS composition stays plastic for a longer period that would help in making a smoother finish for the contractor.
- The sustainability factor of GGBFS is proved from the above-mentioned advantages. The product itself is a by-product that is used for a greater recognition. The reduction of carbon dioxide in the structure make them more sustainable.
- The GGBS as a replacement has lesser water demand because of their glassy texture. The glassy surface of GGBS particles does not absorb water onto its surface.
- GGBS is facing certain hindrances in its implementation widely due to the low rate of strength gain. It is also highly sensitive to conditions of curing. So, a better quality of curing process ensures the happening of full advantage of GGBS.

Composition of Ground Granulated Blast-furnace Slag (GGBS)

- The difference in mineralogical composition in GGBFS compared to Portland cement is shown in the table below.

Material	GGBFS	Portland Cement
CaO	30 - 50%	55 - 66%
SiO_2	28 - 40%	20 - 24%
Al_2O_3	8 - 24%	0 - 8%
MgO	1 - 18%	5%

- Another significant difference in GGBS is that some of the metals within it have not undergone complete oxidation. This change will be reflected in the structural formation in concrete when compared with Portland cement.
- From the above two differences in composition compared with the Portland cement, would show changes in the hydration reaction as well as the hydration products that are given GGBS.

Physical Properties of Ground Granulated Blast-furnace Slag (GGBS)

- Ground granulated blast furnace slags have a color nearly white in appearance. The figure below shows GGBS and ordinary Portland cement, to clearly show the color difference between them.
- Hence the use of GGBS in concrete manufacture would give a light and brighter color to concrete unlike the dark grey color of ordinary cement concrete structures. This is an added advantage of GGBS in terms of aesthetics.

Particle Size Distribution of Ground Granulated Blast-furnace Slag

- It is observed that the particle size of GGBS varies with the grinding techniques used during its manufacture. It is also observed that the product from ball mill has wide particles and narrow particles for the one grounded in airflow mill. Those grounded in vibro-mill have spherical shaped particles.

Using GGBS cement

- In the concrete manufacturing plant, the GGBS can be added along with the Portland cement, water and aggregates. The normal ratio of the mixture remains the same. The studies show that the GGBS can be replaced from 30 to 85 % of the cement weight. Most of the instances we replace 40 to 50%.

Advantages of GGBS in concrete

The incorporation of ground granulated blast-furnace slag in concrete manufacture gains many advantages which are mentioned below:

- GGBS in concrete increases the strength and durability of the concrete structure.
- It reduces voids in concrete hence reducing permeability

- GGBS gives a workable mix.
- It possesses good pumpable and compaction characteristics
- The structure made of GGBS constituents help in increasing sulphate attack resistance.
- The penetration of chloride can be decreased.
- The heat of hydration is less compared to conventional mix hydration.
- The alkali-silica reaction is resisted highly.
- These make the concrete more chemically stable
- Gives good surface finish and improves aesthetics
- The color is more even and light.
- Lower chances of efflorescence
- The maintenance and repair cost of structures are reduced thus increasing the life cycle of concrete structures.
- Unlike cement, GGBS does not produce carbon dioxide, sulphur dioxide or nitrogen oxides.

2.4.2 Aero-gel Insulation

- Silica aerogel is a lightweight solid derived from gel in which the liquid component of the gel has been replaced with gas. These characteristics make aerogels the world's lowest density solid and most effective thermal insulator.
- Contractors are using the new aerogels to insulate homes that are sealed from the outside, both over masonry and under shingles. On wood frame homes, thin strips of aerogel can be applied to studs to prevent thermal bridging, where heat escapes through the walls framing.
- Among the most important parameters of thermal insulation materials are their thermal properties, and in addition their safety in case of fire.

Aerogel insulation for buildings

- Aerogels are synthetic low-density materials with unique physical properties. They are formed by removing the liquid from a gel under special drying conditions, bypassing the shrinkage and cracking experienced during ambient evaporation. This creates a solid three-dimensional nanoporous structure containing 80-99% air.
- Due to their high porosity, aerogels exhibit the lowest thermal conductivity of any solid, whilst being transparent to light and solar radiation. Aerogels are often cited as a promising material for translucent insulation applications. They can made from practically any material, although the most common form is silica aerogel which can be produced as granules or in solid (monolithic) tiles.

Commercial products for the building sector include:

- Cavity insulation.
- Glazing units and cladding systems containing granular aerogel.
- Translucent and opaque insulation boards, blankets and tensile roof membranes embedded with aerogel particles.
- Transparent monolithic silica aerogel has been cited as the 'holy grail' of future glazing technology, with the potential to achieve U-values as low as 0.1 $W/m^2.K$. However, research and development into monolithic glazing is limited due to the high cost of production, long processing time and the difficulty of creating large uniform samples with complete transparency.

2.4.3 Cooling Bricks

- 'Cooling bricks' uses the process of evaporative cooling to cool buildings during hot and dry climates. Cooling Bricks masonry walls can be filled with water just like a sponge which passively cools the interiors of our homes.
- Summers are the time when we depend on air conditioners or air coolers to go through our day and night. A summer day when we might have a power breakdown is like a nightmare. But imagine a building system that can cool our homes using the most abundant compound available on earth – water.

Evaporative Cooling

- Evaporative cooling is a process that can be used to cool buildings. This can be as simple as including a water feature such as a fountain in a courtyard or spraying water over a building roof. It is basically a process of addition of water vapor into the air, which causes a lowering of the ambient temperature.
- The concept of evaporative cooling comes from our body sweating process. When sweat evaporates from our skin, the heat of vaporization is drawn from the skin, which results in cooling of our body temperatures.

The Principle behind Cooling Bricks

- Cooling Bricks masonry walls use this principle to passively cool the interiors of our homes. Made-up of 3 dimensional printed porous ceramic bricks set in mortar, each brick absorbs water like a sponge. These bricks can be filled with water just like a sponge.
- When the hot and dry air passes through the bricks, the air absorbs the water through evaporation and becomes cooler. This cooler air enters our rooms, lowering the interior temperature.
- Cooling Bricks are modular and interlocking. They can be stacked together to create large screens. The 3-dimensional lattice creates a strong bond when set in mortar. The shape of the brick also creates a shaded surface on the wall to keep a large percentage of the wall's surface cool and protected from the sun. This further improves the wall's performance.
- For **Emerging Objects,** all of their focus is around developing 3D printing solutions and designing innovative spaces for living and working. A number of their innovations are still at the prototype stage.
- According to the design and engineering teams at **Emerging Objects** if you live in a hot and dry climate such as the Middle East and Western Australia you can construct an entire wall or home using Cooling Bricks which will act as your cooling system. They will work wonders in such regions by providing for the capability of holding a lot of water vapour.
- So, instead of running air containers and incurring huge electricity bills, you could just water your walls and enjoy a pleasant and cool temperature inside your house. This could bring a major impact in the desert environment though it would not be of much use in areas where humidity levels are already high.
- In my opinion, while this is an innovative design and future technology, inherently Cooling Bricks are ceramic based and hence they would lack the toughness and security that regular bricks give our homes. They may also not be appropriate for internal walls because of their brittle nature.
- Having said that, the power of this new technology and innovation can be harnessed as a design element to our walls. Interspersing Cooling Bricks between regular bricks or having a portion of the wall to stand out aesthetically. This can still provide for the cooling effect of the bricks and not compromise on safety and security.
- All in all, Cooling Bricks is an exciting innovation in building material and it would be interesting to see how **Emerging Objects** makes it more practical.

2.4.4 Green Concrete

- Concrete which is made from concrete wastes that are eco-friendly are called as "Green concrete".
- Green Concrete is a term given to a concrete that has had extra steps taken in the mix design and placement to insure a sustainable structure and a long life cycle with a low maintenance surface. For example, Energy saving, CO_2 emissions, wastewater.
- Today the word green is not just limited to colour, it represents the environment, which is surrounding us.
- "Green concrete" is a revolutionary topic in the history of concrete industry. Concrete wastes like slag, power plant wastes, recycled concrete, mining and quarrying wastes, waste glass, incinerator residue, red mud, burnt clay, sawdust, combustor ash and foundry sand.

- The goal of the **Centre for Green Concrete** is to reduce the environmental impact of concrete. To enable this, new technology is developed. The technology considers all phases of a concrete construction's life cycle, i.e. structural design, specification, manufacturing and maintenance, and it includes all aspects of performance, i.e.
    1. Mechanical properties (strength, shrinkage, creep, static behaviour etc.).
    2. Fire resistance (spalling, heat transfer etc.).
    3. Workmanship (workability, strength development, curing etc.).
    4. Durability (corrosion protection, frost, new deterioration mechanisms etc.).
    5. Thermodynamic properties (input to the other properties).
    6. Environmental aspects (CO_2-emission, energy, recycling etc.).

There are a number of alternative environmental requirements with which green concrete structures must comply:

- CO_2 emissions shall be reduced by at least 30 %.
- At least 20 % of the concrete shall be residual products used as aggregate.
- Use of concrete industries own residual products.
- Use of new types of residual products, previously land filled or disposed of in other ways.
- CO_2-neutral, waste-derived fuels shall substitute fossil fuels in the cement production by at least 10 %.

Advantages of Green Concrete

- Reduction of the concrete industry's CO_2-emmision by 30 %.
- Increased concrete industry's use of waste products by 20%.
- No environmental pollution and sustainable development.
- Green concrete requires less maintenance and repairs.
- Green concrete having better workability than conventional concrete.
- Good thermal resistant and fire resistant.
- Compressive strength behaviour of ceracrete with water cement ratio is similar to conventional concrete.
- Flexural strength of green concrete is almost equal to that of conventional concrete.

Limitations of Green Concrete

- By using stainless steel, cost of reinforcement increases.
- Structures constructed with green concrete have comparatively less life than structures with conventional concrete.
- Split tension of green concrete is less than that of conventional concrete.

2.4.5 Timbercrete

- Timbercrete is made of a unique blend of cellulose (timber waste), cement, sand, binders, and other materials.
- Timbercrete is a blend of sawmill waste, cement, sand, binders and a non-toxic deflocculating additive, which is cured using the renewable resources of sun and wind into a unique building block.
- It is then converted into bricks, blocks, panels and pavers that are used not only in residential, industrial and commercial building projects, but also landscaping design.
- It can be moulded into a vast array of different sizes, shapes, colours and textures, with a few qualities that make it a true standout when compared to conventional bricks or cement.
- It's around 2.5 times lighter than concrete or clay and has a semi-flexible quality that improves its engineering diversity and application, together with the ability to be nailed or screwed just like timber.

- Compared to clay and concrete products, it has an improved insulation value and thermal mass, meaning that it will store thermal energy more efficiently and release it more slowly, reducing heating and air-conditioning costs and the environmental knock-on effects.
- Its fire rating is greatly enhanced compared to concrete, clay, timber or steel and it's even bullet proof against a huge array of weaponry.
- Timbercrete has been tested to ensure that it is completely safe and non-toxic, ensuring that it's suitable for those sensitive to allergens.
- Timbercrete has been used as a resource in building homes for under privileged families through the Habitat for Humanity organization, as well as training indigenous communities to manufacture blocks for their own housing use. Micro-enterprises and training programs have been set-up in India.

Recycled Waste:

- As the name suggests "Timber-crete" is made primarily from timber waste such as sawdust or recycled timber from discarded pallets and the like. No trees have been cut down specifically to produce Timbercrete. Instead, we use the excess sawmill waste from plantation timbers that would otherwise be burnt or used in a form that would cause it to degrade, releasing greenhouse gases into our atmosphere.

Other Ingredients

- Sand - Sand is carefully selected (sometimes blended) to maximise its load bearing capacity (MPA) and minimise water ingress.
- Binders - Cement binders, such as Portland cement or other cementitious material, along with a special nontoxic "deflocculate" and other products that improve density and cement performance. These products also aid as a waterproofing agent.

2.4.6 Ferrock

- The name Ferrock is a reflection of its composition – largely iron-rich ferrous rock.
- It's actually created from waste steel dust which is normally discarded from industrial processes and silica from ground up glass.
- While looking into how to keep iron from rusting and hardening up, the material that inspired Ferrock was unintentionally created. Stone initially discarded this incredibly hard material, then suddenly realized it might have a useful purpose as a more environmentally friendly, yet robust, alternative to cement.
- In the experiments which followed, his aim was to find a material that could be mixed and poured just like cement and with the same strength and versatility that concrete offers.
- The iron within the steel dust reacts with CO_2 and rusts to form iron carbonate. It's this that is fused into the matrix of Ferrock and, like concrete, after it's dried, it cannot be melted back into a liquid form but retains its hard, rock-like qualities.
- Even if Ferrock is not the exact answer or a cost-effective solution in the long-term future, it's a promising starting point for developing smarter technologies that address our insatiable hunger for development and the devastating result it's having on our environment.

Advantages

- Compared to Portland cement (made from chalk and clay and resembling Portland stone in color), which is one of the leading types in use throughout the world today, Ferrock is actually five times stronger.
- It can withstand more compression before breaking and is far more flexible, meaning it could potentially resist the earth movements caused by seismic activity or industrial processes.
- One of the unique properties of Ferrock is that it becomes even stronger in salt water environments, making it ideal for marine-based construction projects.
- And rather than emitting large amounts of CO_2 as it dries, Ferrock actually absorbs and binds it! This results in a carbon-negative process that actually helps to trap greenhouse gases.

Important Points

- Sand manufactured by Vertical Shaft Impactor is of cubical shape. Such sand can be used for all types of construction work, Concreting, Plastering etc and is better substitute to river sand. V.S.I. Crushers is a most economical machine for Crushing Stone in Cubical shape and manufacturing artificial sand.
- The most common component of sand is silicon dioxide in the form of quartz. The Earth's landmasses are made up of rocks and minerals, including quartz, feldspar and mica. Weathering processes – such as wind, rain and freezing/thawing cycles – break down these rocks and minerals into smaller grains.
- Manufactured sand (M-Sand) is a substitute of river sand for concrete construction. Manufactured sand is produced from hard granite stone by crushing. The crushed sand is of cubical shape with grounded edges, washed and graded to as a construction material. The size of manufactured sand (M-Sand) is less than 4.75mm.
- A fine concrete mix embedded with light conducting channels and formed into slabs or tiles that shift, shimmer, and twinkle in response to light, movement and shadows.
- Carbon fibers or carbon fibres are fibers about 5–10 micrometres in diameter and composed mostly of carbon atoms. Carbon fibers have several advantages including high stiffness, high tensile strength, low weight, high chemical resistance, high temperature tolerance and low thermal expansion.
- Carbon fiber is made from organic polymers, which consist of long strings of molecules held together by carbon atoms. Most carbon fibers (about 90 percent) are made from the polyacrylonitrile (PAN) process. A small amount (about 10 percent) is manufactured from rayon or the petroleum pitch process.
- Carbon fiber has gone to the moon on spacecraft, but it is also used widely in aircraft components and structures, where its superior strength to weight ratio far exceeds that of any metal. 30 percent of all carbon fiber is used in the aerospace industry.
- Cigarette filters are made of a plastic called cellulose acetate. When tossed into the environment, they dump not only that plastic, but also the nicotine, heavy metals, and many other chemicals they've absorbed into the surrounding environment.
- Cigarette butt bricks are produced by blending cigarette butts with mixture of materials used for brick production. This is an innovative and practical solution for the pollution problems caused by cigarette butt disposal.
- 3D Printed Bricks are be the building blocks of modern, sustainable architecture people have built homes by shaping mud into blocks, drying them in the sun, and stacking them. Using clay and sand, 3D print bricks are designed blocks that can cool down rooms and even stabilize buildings when earthquakes strike. They are also working with unconventional materials such as salt and sawdust-techniques that could one day allow waste to be transformed into dwellings.
- The Cool Brick is a 3D-printed ceramic object about six inches long, shaped like an elongated hexagon. In addition to being very porous, the brick resembles a small lattice with larger holes that air can blow through. When water is dripped onto the wall with a plumbing system or splashed on by hand, it is absorbed into the ceramic and then evaporates, cooling the air. Breezes blowing through the holes bring cool, humid air into the room.
- Transparent wood composites are novel wood materials which have up to 90% transparency and higher mechanical properties than wood itself, made for the first time in 1992. When these materials are commercially available, a significant benefit is expected due to their inherent biodegradable properties since it is wood.
- Transparent wood is one of the innovative construction materials that hold the potential to lift the construction sector to a whole new level. Not only is it environmentally-friendly, but also a great alternative to plastic and glass. Transparent wood is produced by treating and compressing wood strips.

- Glued laminated timber, also abbreviated glulam, is a type of structural engineered wood product constituted by layers of dimensional lumber bonded together with durable, moisture-resistant structural adhesives.
- Glued timber is a wood product which does not fulfill the laminated timber standards, but is made by gluing two or more items of sawn goods together. Such products are used, as they are, in load bearing construction and as planks for various wood products such as window frames, laminated logs, etc.
- 3D Tiles are an open specification for streaming massive heterogeneous 3D geospatial datasets.
- 3D Tiles are used to stream 3D content, including buildings, trees, point clouds, and vector data on flooring tiles, roofing tiles and wall tiles. These tiles gives a decorative look to the building units.
- Geosynthetics are synthetic products used to stabilize terrain. They are generally polymeric products used to solve civil engineering problems. This includes eight main product categories: geotextiles, geogrids, geonets, geomembranes, geosynthetic clay liners, geofoam, geocells and geocomposites.
- As per the definition of American Society for Testing and Materials (ASTM), a geosynthetic material is a planar product manufactured from a polymeric material that is used with soil, rock, earth, or other geotechnical related material as an integral part of a civil engineering project, structure, or system.
- There are six families of geosynthetics: geotextiles, geogrids, geonets, geomembranes, geocomposites and "geo-others." Geotextiles form the largest group of geosynthetics. They are indeed textiles in the traditional sense, but consist of synthetic fibers rather than natural ones, such as cotton, wool or silk.
- Quiet asphalt pavement can help reduce highway noise by as much as 7 decibels, according to a study. Reducing noise by just 3 decibels is equivalent to doubling the distance from the source of noise to the listener, according to Asphalt Pavement literature.
- Quiet asphalt pavement options fall into four general categories. These are fine-graded surfaces, open-graded friction courses, rubberized asphalt and stone-matrix asphalt.
- Fine-graded surfaces are conventional asphalt mixes that are mostly composed of fine-grained particles; reducing the apparent road noise generated by the expelled air by altering its frequencies. Open-graded friction courses (OGFC) are a porous mix; with more connected voids through which air and water can escape. Rubberized asphalt contain crumbs of recycled tires that provide some flexing in the road surface as tires pass over it, allowing the air a bit more time to be forced out at lower pressure. Stone-matrix asphalt mainly uses coarse aggregate to create a stone skeleton, tied together by asphalt binders and fibers.
- Permeable pavement is a porous urban surface composed of open pore pavers, concrete, or asphalt with an underlying stone reservoir.
- Permeable pavement catches precipitation and surface runoff, storing it in the reservoir while slowly allowing it to infiltrate into the soil below or discharge via a drain tile.
- Permeable pavement is a porous urban surface composed of open pore pavers, concrete, or asphalt with an underlying stone reservoir.
- Permeable pavement catches precipitation and surface run-off, storing it in the reservoir while slowly allowing it to infiltrate into the soil below or discharge via a drain tile.
- Permeable paving is a method of paving vehicle and pedestrian pathways to enable infiltration of storm water run-off. Permeable pavement surfaces typically include pervious concrete, porous asphalt, paving stones and interlocking pavers.
- Plastic roads are made entirely of plastic or of composites of plastic with other materials. Plastic roads are different from standard roads in the respect that standard roads are made from asphalt concrete, which consists of mineral aggregates and asphalt.

- Plastic-bitumen composite roads have better wear resistance than standard asphalt concrete roads. They do not absorb water, have better flexibility which results in less rutting and less need for repair. Road surfaces remain smooth, are lower maintenance, and absorb sound better.
- Solar Roadways (SR) is a modular system of specially engineered solar panels that can be walked and driven upon. Our panels contain LED lights to create lines and signage without paint. They contain heating elements to prevent snow and ice accumulation.
- The applications for solar panels are growing exponentially. Photovoltaic cells work by using semiconductors to absorb light and create a flow of electrons, which can power any number of electrical devices. Put the two together and you get a unique solution: solar panels on our highways.
- Solar Roadway panels are made of tempered (safety) glass. Glass was chosen for its hardness, strength, durability, and transmittance. SR glass is textured to create proper traction for vehicles and pedestrians.
- Anti-icing is a pro-active approach to winter road maintenance. It forms a bond-breaker between the pavement surface and the snow and ice layer which melts snow more quickly and reduces the chance that ice will form and bond to the surface. Sodium chloride (salt), magnesium chloride, calcium chloride, calcium magnesium acetate and potassium acetate are chemicals used to prevent and remove snow and ice from roadways. VDOT uses liquid magnesium chloride, calcium chloride and sodium chloride for anti-icing and pre-treatment.
- Piezoelectric devices, used for harvesting the vibrational energy of roads and walkways due to traffic, can produce electrical energy that is predictable (based on traffic patterns), and locally storable. Piezoelectric devices generate electrical energy by means of a piezoelectric crystal electric floors are designed to capture the wasted energy and resources, and store or redistribute them where they are needed. Energy is generated when a person steps on tiles that feature piezoelectric attributes. This kinetic energy is converted into electricity.
- Portland Pozzolana Cement is a kind of Blended Cement which is produced by either intergrinding of OPC clinker along with gypsum and pozzolanic materials in certain proportions or grinding the OPC clinker, gypsum and Pozzolanic materials separately and thoroughly blending them in certain proportions.
- PORTLAND POZZOLANA CEMENT IS Used in hydraulic structures, marine structures, construction near the sea shore, dam construction etc. Used in pre-stressed and post-tensioned concrete members. As it gives better surface finish, it is used in decorative and art structures.
- PPC offers many other advantages, such as durability. 6. Portland pozzolana cement improves pore size distribution and also reduces the micro cracks in the cement paste at the transition zone due to pozzolanic action and being finer than OPC.
- The Masterglanium product comprises new generation high-range water-reducing admixtures that are specially formulated for applications where slump VAUE OF CONCRETE retention, high/early strengths and durability. Concrete mixtures containing these premier products can be optimized for delivery in remote locations and for use in hot and cold climates.
- The technology and versatility of the Masterglanium product portfolio excels in high performance areas such as high strength concrete, pumped CONCRETE self-consolidating concrete, low permeability, low water absorption concrete and low shrinkage concrete.. Its technology provides excellent concrete performance in both the fresh and hardened states with good surface appearances.
- Based on the latest generation of polymer technology, the Masterglanium admixture is made up of many customized molecules.
- Polycarboxylic Ether is a second invention compound, which produce very strong and high gains. PCE-811 is the best solution for creating the precast solid constituents along with all procedures, including Rheoplastic or Super feasible concrete.

- Concrete plasticizer which is commonly known as water reducing admixture amongst professionals is used for improving workability of concrete. ... Concrete plasticizers increases the workability of freshly mixed concrete without increasing the water cement ratio.
- Self-healing concrete is a product that will biologically produce limestone to heal cracks that appear on the surface of concrete structures.
- This idea of self healing concrete has been brought about by a scientist Hendrik Jonker. A bacteria Bacillus pseudofirmus use calcium lactate as afood source producing limestone as a by product. It is said that they can be placed either in small capsules with the nutrient (to keep the bacteria dormant mixed)in the concrete.
- While making self healing concrete Specially selected types of the bacteria genus Bacillus, along with a calcium-based nutrient known as calcium lactate, and nitrogen and phosphorus, are added to the ingredients of the concrete when it is being mixed. These self-healing agents can lie dormant within the concrete for up to 200 years.
- Fiber-reinforced concrete is concrete containing fibrous material which increases its structural integrity. It contains short discrete fibers that are uniformly distributed and randomly oriented.
- Fibers also lower the permeability of concrete and thus reduce bleeding of water. Some types of fibres produce greater impact, abrasion and shatter resistance in concrete. Generally fibres do not increase the flexural strength of concrete, so it can not replace moment resisting or structural steel reinforcement.
- High-performance means that the concrete has one or more of the following properties: low shrinkage, low permeability, a high modulus of elasticity, or high strength. ... High-strength concrete is typically recognized as concrete with a 28-day cylinder compressive strength greater than 6000 psi or 42 Mpa.
- High-strength concrete is typically used in the erection of high-rise structures. It has been used in components such as columns (especially on lower floors where the loads will be greatest), shear walls, and foundations. High strengths are also occasionally used in bridge applications as well.
- Nano concrete is defined as a concrete made by filling the pores in traditional concrete using nano particles of size < 500 nano meters. ... Then the mechanical properties of concrete such as compressive strength, tensile strength and flexural strength of the respective specimens were tested after 7 days and 28 days curing.
- Light transmitting concrete also known as Translucent concrete is a concrete based building material with light-transmissive properties due to embedded light optical elements — usually optical fibers. Light is conducted through the stone from one end to the other. Therefore, the fibers have to go through the whole object.
- Concrete you can see through. The translucent material is created by combining concrete and thousands of optical fiber strands that act like aggregate. The optical fibers form a matrix between the two main surfaces of concrete block, connecting and directing light between the two block faces.
- The ground granulated blast furnace slag (GGBFS) is a by-product of iron manufacturing which when added to concrete improves its properties such as workability, strength and durability. ... This slag is later granulated by cooling it. For this, it is allowed to pass through a high-pressure water get.
- It is a granular product with very limited crystal formation, is highly cementitious in nature and, ground to cement fineness, and hydrates like Portland cement. GGBS is used to make durable concrete structures in combination with ordinary Portland cement and/or other pozzolanic materials.
- The reaction heat per unit mass of GGBS is only 16. 7% higher than that of fly ash. The main reason for the difference of hydration heats between GGBS binder and fly ash binder is that GGBS shows much higher hydraulic activity than fly ash.
- Silica aerogel is a lightweight solid derived from gel in which the liquid component of the gel has been replaced with gas. These characteristics make aerogels the world's lowest density solid and most effective thermal insulator.

- Contractors are using the new aerogels to insulate homes that are sealed from the outside, both over masonry and under shingles. On wood frame homes, thin strips of aerogel can be applied to studs to prevent thermal bridging, where heat escapes through the walls framing.
- Cooling bricks are used for Evaporative cooling is the addition of water vapor into air, which causes a lowering of the temperature of the air. Before the advent of refrigeration, evaporative cooling was used for millennia. Porous ceramic vessels were used to cool water by evaporation through their walls.
- Green concrete is defined as a concrete which uses waste material as at least one of its components, or its production process does not lead to environmental destruction, or it has high performance and life cycle sustainability.
- Green concrete capable for sustainable development is characterized by application of industrial wastes to reduce consumption of natural resources and energy and pollution of the environment. Marble sludge powder can be used as filler and helps to reduce the total voids content in concrete.
- Timbercrete is a multi-award winning and environmentally sustainable masonry building product. We make bricks, pavers and cladding.

Multiple Choice Questions

1. Sand is used for ......
   (a) plastering material (b) flooring tile fixing material
   (c) concrete making (d) all of the above
2. Weathering of rocks takes place due to ....
   (a) wind (b) river
   (c) rain (d) all of these
3. "M-Sand" is known as ......
   (a) market sand (b) moulded sand
   (c) manufactured sand (d) mobile sand
4. Diameter of carbon fiber is form ......
   (a) 5 cm to 10 cm (b) 5 mm to 10 mm
   (c) 5 micrometer to 10 micrometer (d) none of these
5. Cigarette filter waste gives us ......
   (a) cement (b) plastic
   (c) ash (d) carbon
6. 3-d-printed bricks are made from ......
   (a) plastic (b) glass
   (c) clay and sand (d) concrete
7. Transparent wood is made from ......
   (a) plastic (b) glass
   (c) rubber (d) wood strips
8. Glued timber is a type of product made from ......
   (a) rubber and glue (b) glass and glue
   (c) timber and glue (d) plastic and glue
9. Geosynthetic materials are used to strengthen ......
   (a) soil (b) concrete
   (c) glass (d) brick wall
10. Which is not a type of geosynthetic material ?
    (a) Geotextile (b) Geogrides
    (c) Geonets (d) Geoconcrete

11. Quiet asphalt pavement helps in reducing ......
    (a) noise on highway
    (b) less friction on wheels on highway
    (c) initially cost of pavement
    (d) maintenance cost of pavement
12. Permeable pavement allows ......
    (a) vehicle to move smoothly
    (b) rain water to percolate in ground below pavement
    (c) rain water to accumulate on pavement
    (d) dust control on pavement
13. Plastic roads are best suited ......
    (a) in area of heavy rainfall
    (b) in area of low rainfall
    (c) in areas of high temperature
    (d) none of these
14. LED lights are used in construction of ......
    (a) solar highways
    (b) plastic roads
    (c) paver blocks
    (d) gutter on highways
15. A bond between road surface and ice is broken by ......
    (a) concrete
    (b) AntiIcing material
    (c) Ice
    (d) none of these
16. Piezoelectric devices produces ......
    (a) electricity due to vibration of road surface
    (b) vibrations on road surface
    (c) stability of road surface
    (d) anti skid property on road surface
17. Cement used for hydraulic structures and marine structure is ....
    (a) ordinary Portland cement
    (b) Pozzolana Portland cement
    (c) white cement
    (d) oil well cement
18. Master glenium is a type of ....
    (a) water reducing admixture in concrete
    (b) slump value accelerator in concrete
    (c) slump value retarder in concrete
    (d) early strength admixture in concrete
19. "PCE" is a material and known as ......
    (a) poly carbon ethylene
    (b) poly carboxylic ether
    (c) poly centric ether
    (d) poly concrete ethanol
20. Self healing concrete comprises of ......
    (a) a bacteria Bacillus pseudofirmus
    (b) special sand
    (c) coloured pigments
    (d) ice and cement
21. Fiber reinforced concrete is prepared by mixing which ingredient along with cement, sand, coarse aggregate and water ......
    (a) asphalt fibers
    (b) steel fibers
    (c) vegetable fibers
    (d) brick bats
22. A concrete having pores filled with particles of very small size of 5×10^{-7} meter is called ......
    (a) micro concrete
    (b) macro concrete
    (c) nano concrete
    (d) none of these
23. Concrete you can see through is ......
    (a) partial concrete
    (b) translucent concrete
    (c) high performance concrete
    (d) ground water concrete

Answers

1. (d)	2. (d)	3. (c)	4. (c)	5. (b)	6. (c)	7. (d)	8. (c)	9. (a)	10. (d)
11. (a)	12. (c)	13. (a)	14. (a)	15. (b)	16. (a)	17. (b)	18. (a)	19. (b)	20. (a)
21. (c)	22. (c)	23. (b)							

❑❑❑

Chapter **3**

Latest Tools and Equipments

Weightage of Marks = 14, Teaching Hours = 10

Syllabus

3.1 Survey equipment

- LiDAR, Scan Station, Global Positioning System (GPS), Geographical Information System (GIS), Photogrammetry, Drones, Direct Reading Grade Rods, 3D Laser scanning, laser level.

3.2 Construction Equipment

- Earth moving equipment- Skid and crawler loaders, trenchers, scrappers, wheeled loading shovels, advanced plastering machine, Bridge launcher.

3.3 Material handling equipment

- Cranes, conveyors, hoists, forklifts, mobile concrete mixer, paver, road header, tunnel boring machine.

About this Chapter

At the end of this chapter, students will be able to:

- Identify the correct use/s of given advanced tool/equipment.
- Select the appropriate instrument required for the given construction activity.
- Identify the situation for the use of given advance equipment/instrument.

3.1 SURVEY EQUIPMENT

- Modern surveying instruments provides faster and more precise surveying than conventional instruments.
- In conventional surveying, chain and tape are used for making linear measurements while compass and ordinary theodolites are used for making angular measurements. Leveling work is carried out using a Dumpy level and a levelling staff. With such surveying instruments, survey work will be slow and tedious.
- Hence modern surveying instruments are becoming more popular and they are gradually replacing old surveying instruments such as compass and Dumpy level. With modern surveying instruments, survey work will be precise, faster and less tedious.

3.1.1 LiDAR

- "LIDAR, which stands for Light Detection and Ranging, is a remote sensing method that uses light in the form of a pulsed laser to measure ranges (variable distances) to the Earth. These light pulses - combined with other data recorded by the airborne system - generate precise, three-dimensional information about the shape of the Earth and its surface characteristics.
- A LIDAR instrument principally consists of a laser, a scanner, and a specialized GPS receiver. Airplanes and helicopters are the most commonly used platforms for acquiring LIDAR data over broad areas. Two types of LIDAR are topographic and bathymetric. Topographic LIDAR typically uses a near-infrared laser to map the land, while bathymetric lidar uses water-penetrating green light to also measure seafloor and riverbed elevations.
- LiDAR, which stands for Light Detection and Ranging, is a remote sensing method that uses light in the form of a pulsed laser to measure ranges (variable distances) to the Earth.

- As it turns out, the accuracy of the laser ranging is actually one of the most accurate components in the whole system.
- Although airborne LiDAR systems may not be this accurate, as it turns out the error contribution from the laser ranging is by far one of the least significant sources of overall error.
- LiDAR is a surveying method that measures distance to a target by illuminating the target with laser light and measuring the reflected light with a sensor. Differences in laser return times and wavelengths can then be used to make digital 3-D representations of the target.
- The days of *feet on the ground* surveying with conventional surveying equipment (GPS, total stations, levels) for topo collection of 5 to several thousand acres is fast becoming a thing of the past. With a Revolution or A-series system thousands of acres can easily be scanned in a day. The surveyor still needs to be present to locate sub-surface utilities but all visible information can be captured with the LiDAR and extracted in the office.
- The least expensive way to get started is to use a ScanLook Revolution. It's small, light, and affordable. It's geo-positioning uses only L1 SBAS so you cannot get centimeter level results directly but this can be improved using a few control points or an external positioning device (i.e. re-use your GPS equipment).
- If you can afford a bit more money up front and want to avoid using control as frequently, then the ScanLook A-Series is right for the job. It can get centimeter level results using only $\frac{L_1}{L_2}$ GNSS and post processing.
- Combine affordable LiDAR capture with some good software for bare earth classification to remove the vegetation and you have a topographic mapping system that will change your business model.

3.1.2 Scan Station

- For many as-built and topographic surveys, laser scanners have already proven to significantly reduce field labor compared to traditional methods.
- Scan Station 2's jump-step increase in pulsed scan speed further slashes these costs – field labor is now as little as one-fifth of that of traditional methods for many projects.
- It adds long range scanning of more than 1 kilometre to the well-known features such as high speed, excellent data quality, HDR imaging, Check and Adjust, ease-of-use and ruggedness.

Dramatically faster scanning also lets users:

- Collect data in tighter time windows.
- Reduce time spent in hazardous locations.
- Provide project results faster.
- Collect even more complete data.
- "Squeeze in" additional service requests from clients.

3.1.3 Global Positioning System (G.P.S.)

- G.P.S., which stands for Global Positioning System, is a radio navigation system that allows land, sea, and airborne users to determine their exact location, velocity, and time 24 hours a day, in all weather conditions, anywhere in the world.
- G.P.S. systems are extremely versatile and can be found in almost any industry sector. They can be used to map forests, help farmers harvest their fields, and navigate airplanes on the ground or in the air. G.P.S. systems are used in military applications and by emergency crews to locate people in need of assistance.
- Global Positioning System (G.P.S.) is developed by U.S. Defense department and is called Navigational System with Time and Ranging Global Positioning System (NAVSTAR G.P.S.) or simply G.P.S.
- For this purpose U.S. Air Force has stationed 24 satellites at an altitude of 20200 km above the earth's surface. The satellites have been positioned in such a way, at least four satellites will be visible from any point on earth.

- The user needs a G.P.S. receiver to locate the position of any point on ground. The receive processes the signals received from the satellite and compute the position (latitude and longitude) and elevation of a point with reference to datum.
- The surveying and mapping community was one of the first to take advantage of G.P.S. because it dramatically increased productivity and resulted in more accurate and reliable data. Today, G.P.S. is a vital part of surveying and mapping activities around the world.
- When used by skilled professionals, G.P.S. provides surveying and mapping data of the highest accuracy. G.P.S.-based data collection is much faster than conventional surveying and mapping techniques, reducing the amount of equipment and labor required. A single surveyor can now accomplish in one day what once took an entire team week to do.
- G.P.S. supports the accurate mapping and modeling of the physical world — from mountains and rivers to streets and buildings to utility lines and other resources. Features measured with G.P.S. can be displayed on maps and in geographic information systems (G.I.S.) that store, manipulate, and display geographically referenced data.
- Governments, scientific organizations, and commercial operations throughout the world use G.P.S. and G.I.S. technology to facilitate timely decisions and wise use of resources. Any organization or agency that requires accurate location information about its assets can benefit from the efficiency and productivity provided by G.P.S. positioning.
- Unlike conventional techniques, G.P.S. surveying is not bound by constraints such as line-of-sight visibility between survey stations. The stations can be deployed at greater distances from each other and can operate anywhere with a good view of the sky, rather than being confined to remote hilltops as previously required.
- G.P.S. is especially useful in surveying coasts and waterways, where there are few land-based reference points. Survey vessels combine G.P.S. positions with sonar depth soundings to make the nautical charts that alert mariners to changing water depths and underwater hazards. Bridge builders and offshore oil rigs also depend on G.P.S. for accurate hydrographic surveys.
- Land surveyors and mappers can carry G.P.S. systems in backpacks or mount them on vehicles to allow rapid, accurate data collection. Some of these systems communicate wirelessly with reference receivers to deliver continuous, real-time, centimeter-level accuracy and unprecedented productivity gains.

3.1.4 Geographical Information System (G.I.S.)

- A Geographic Information System is a system designed to capture, store, manipulate, analyze, manage, and present spatial or geographic data. G.I.S. applications are tools that allow users to create interactive queries, analyze spatial information, edit data in maps, and present the results of all these operations.
- The G.I.S. is used to study geologic features, analyze soils and strata, assess seismic information, and or create three dimensional (3D) displays of geographic features. G.I.S. can be also used to analyze rock information characteristics and identifying the best dam site location.
- A Geographic Information System (G.I.S.) is a system designed to capture, store, manipulate, analyze, manage, and present spatial or geographic data. GIS applications are tools that allow users to create interactive queries (user-created searches), analyze spatial information, edit data in maps, and present the results of all these operations.
- G.I.S. can refer to a number of different technologies, processes, techniques and methods. It is attached to many operations and has many applications related to engineering, planning, management, transport/logistics, insurance, telecommunications, and business. For that reason, G.I.S. and location intelligence applications can be the foundation for many location-enabled services that rely on analysis and visualization.

- G.I.S. can relate unrelated information by using location as the key index variable. Locations or extents in the Earth space–time may be recorded as dates/times of occurrence, and x, y and z co-ordinates representing, longitude, latitude, and elevation, respectively. All Earth-based spatial–temporal location and extent references should be relatable to one another and ultimately to a "real" physical location or extent. This key characteristic of GIS has begun to open new avenues of scientific inquiry.
- **Mapping where things are.** We can map the spatial location of real-world features and visualize the spatial relationships among them. For example: below we see a map of frac sand mine locations and sandstone areas in Wisconsin. We can see visual patterns in the data by determining that frac sand mining activity occurs in a region with a specific type of geology.
- **Mapping quantities.** People map quantities, such as where the most and least are, to find places that meet their criteria or to see the relationships between places. For example : below is a map of cemetery locations in Wisconsin. The map shows the cemetery locations as dots (dot density) and each county is color coded to show where the most and least are (lighter blue means fewer cemeteries).
- **Mapping densities.** Sometimes it is more important to map concentrations, or a quantity normalized by area or total number. For example: Below we have mapped the population density of Manhattan (total population counts normalized by the area in sq. miles of census tracts.)
- **Finding what is inside.** We can use G.I.S. to determine what is happening or what features is located inside a specific area/region. We can determine the characteristics of "inside" by creating specific criteria to define an area of interest (AOI). For example: below is a map showing a flood event and the tax parcels and buildings in the floodway. We can use tools like CLIP to determine which parcels fall inside the flood event. Further, we can use attributes of the parcels to determine potential costs of property damage.
- **Finding what is nearby.** We can find out what is happening within a set distance of a feature or event by mapping what is nearby using geo-processing tools like BUFFER. For example: below we see a map of drive times from a central location in the City of Madison, WI. We can use streets as a network and add specific criteria like speed limit and intersection controls to determine how far a driver can typically get in 5, 10, or 15 minutes. (Map courtesy of UW Extension)
- **Mapping change.** We can map the change in a specific geographic area to anticipate future conditions, decide on a course of action, or to evaluate the results of an action or policy. Example: below we see land use maps of Barnstable, MA showing changes in residential development from 1951 to 1999. The dark green shows forest, while bright yellow shows residential development. Applications like this can help inform community planning processes and policies.

3.1.5 Photogrammetry

- Photogrammetry is the science of making measurements from photographs. The input to photogrammetry is photographs, and the output is typically a map, a drawing, a measurement, or a 3D model of some real-world object or scene.
- Photogrammetry has a long history of use in the surveying and mapping field. Photogrammetry can be used for point measurements or for generating point clouds. Compared to its chief rival, LiDAR (Laser Imaging, Detection and Ranging), point-cloud generation from photogrammetry is cost-effective, portable, and versatile, so it has always had a loyal following among practitioners.
- There are two types of Photogrammetry, Aerial Photogrammetry and Terrestrial (Close Range) Photogrammetry.
- Aerial digital photogrammetry, often used in topographical mapping, begins with digital photographs or video taken from a camera mounted on the bottom of an airplane.

- Photogrammetry is the use of photography in surveying and mapping. It's used to measure distances between objects, and photogrammetry software can utilize images to create 3D renderings. While in its early days it was used mostly for topographic mapping, it's recently been used in a number of unexpected ways.
- The Photogrammetry is a surveying and mapping technique which has several applications in the Transportation Department. The practice of Photogrammetry Applications in surveying includes siteplanning, topographic mapping, earthwork volume estimation for proposed roads, a collection of digital elevation models (D.E.M.), and image-based mapping (orthophotography).
- The term "photogrammetry" is the combination of the words "photo" and "meter" it means measurements from photographs. The classical definition of photogrammetry is:
- "It is the art, science, and technology of gathering reliable information regarding any physical object and the complete environment. This technique involves a process of record-keeping, measuring, and evaluating terrestrial and aerial photographs.
- Photogrammetry refers to art because to obtain reliable measurements it requires specific skills, techniques, and decisions taken by an individual. Photogrammetry is a science and a technology because it uses images and modifies it, through technology, into significant results. The Modern photogrammetry involves image references and image forms different than photographs, like Radar images.
- The photogrammetric process consists of proper project planning, image retrieval, image processing, control data for image adjustment, data accumulation and presentation of an end product results. The result of the photogrammetric method co-ordinates values of specific points, a realistic representation of the earth surface (topographic map), or a reformed image of the earth surface including map-like characteristics (orthophoto.)
- Photographs used for photogrammetry are capture from a special (metric) camera, a typical camera or from digital sensors. The images are captured and recorded from a tool fixed on a satellite, on an aircraft as well as helicopters, or on a tripod (terrestrial photogrammetry) which is set-up and mounted on the ground.

Benefits and Drawbacks of Photogrammetry

Here are few Benefits of photogrammetry over standard surveying and mapping methods are as follows:

- While capturing the Aerial photographs, photogrammetry produces an actual and permanent photographic record of a particular condition that exists while capturing images. Since the record has metric properties, it's not only a pictorial record, but also it is an accurate, measured record.
- For information that has to be re-surveyed or re-evaluated, there is no need to perform any expensive fieldwork. The same images we can measure again and can get new information in a very convenient manner. We can quickly remedy the missing information, such as inadequate offsets for cross-sections.
- It provides a wide mapped area so other line studies can perform with the corresponding data source more efficiently than other traditional methods. Photogrammetry presents a broad view of the project field, by classifying both topographic and cultural features.
- It is useful in any locations that are difficult, unsafe, or impossible to access. Photogrammetry is an ideal surveying method for toxic areas where fieldwork may negotiate the safety of the surveying crew.
- A remarkable benefit of photogrammetry is that the road surveys can work without disturbing the traffic by closing lanes, or endanger the field team. Once it captures the roads the analysis of road features, including elevation data, we can do it in the office, not in the actual field.
- The co-ordinates of each point in the mapping field can determine with no additional efforts or cost.
- The aerial images are useful for conveying or describing information to the public, State, including Federal agencies, as well as to other divisions of Transportation.

Here are some drawbacks:

- Climate statuses (winds, clouds, haze etc.) affect the aerial photogrammetry process and the image quality.
- Seasonal states affect the aerial images, i.e., snow cover will defeat the targets and defines a false ground impression. Hence, there is only a short-term period typically November through March that is perfect for common purpose aerial photogrammetry.
- A cleared construction section or a highway that is not blocked by trees is limited to this restriction. These kinds of projects can fly and photograph.
- Hidden areas caused by human-made objects, such as a roof of a house, cannot map with photogrammetry. Unknown area problems can cause by tree canopy, thick vegetation, or by rugged terrain with sharp inclines. The information hidden from the camera must map with other surveying methods.

The Top Components of Photogrammetry:

In simple words, photogrammetry has three essential components. They are image control, image acquisition, and product compilation.

- Image acquisition covers planning the overflight, selecting a relevant camera system, photo capturing film processing, film review and annotation, printing of paper copies, and image scanning (if needed.)
- The Image control component involves the process of selecting areas for ground control and targeting, area surveying control details, and aerial triangulation. This component can get eliminate if advanced GPS methodology can solve the photo orientation difficulty without requiring ground control.
- The product compilation component varies and depends on the quality of the product. Topographic maps, orthophotos, or monoscopic updates are all photogrammetric products which are organized in many ways as explained later. Each one of these components needs the utilization of various equipment, different measurement techniques, and different data processing.

3.1.6 Drones

- A typical unmanned aircraft is made of light composite materials to reduce weight and increase maneuverability. Drones are controlled by remote ground control systems (G.S.C.) and also referred to as a ground cockpit. An unmanned aerial vehicle system has two parts, the drone itself and the control system.
- Drones Aerial survey is a method of collecting geomatices or other imagery by using aero planes, helicopters... etc.
- Drones mean a quantum leap for surveying. They provide eyes that can reach and hover above specific sites. Their height and cameras can be adjusted remotely. And, equipped with sensors, they can measure, transmit and store data.
- Global positioning has created greater interest in Geographical Information Systems (G.I.S.). According to Drone Analyst, "G.I.S. professionals provide a wide variety of land-related services like identifying property boundaries, subdividing land, and surveying construction sites for placement of buildings. They also produce topographic and hydrographic maps, volumetric calculations for stockpiles, and flood insurance maps, among other services."

Benefits of drones in surveying

1. **Reduce field time and survey costs**

- Capturing geospatial data with a drone is up to five times faster than with land-based methods and requires less manpower. With PPK geo-tagging, you also save time, as placing numerous GCPs is no longer necessary. You ultimately deliver your survey results faster than ever and at a lower cost.

2. Provide accurate and exhaustive data

- Total stations only measure individual points. One drone flight produces thousands of measurements, which can be represented in different formats (orthomosaic, point cloud, DTM, DSM, contour lines etc.). Each pixel of the produced map or point of the 3D model contains 3D geo-data.

3. Map otherwise inaccessible areas

- A mapping drone can take off and fly almost anywhere. You are no longer limited by unreachable areas, unsafe steep slopes or harsh terrain unsuitable for traditional measuring tools. You do not need to close down highways or train tracks, but you can capture data during operation without an organizational overhead.

Main applications

- Survey drones generate high-resolution orthomosaics and detailed 3D models of areas where low-quality, outdated or even no data, are available. They thus enable cadastral maps to be produced quickly and easily, even in complex or difficult to access environments. Surveyors can also extract features from the images such as signs, curbs, road markers, fire hydrants and drains.
- After post-processing with a photogrammetry software, these same images can produce very detailed elevation models, contour lines and breaklines, as well as 3D reconstruction of land sites or buildings.
- Aerial images taken by drones greatly accelerate and simplify topographic surveys for land management and planning. This holds true for site scouting, allotment planning and design, as well as final construction of roads, buildings and utilities.
- These images also provide the foundation for detailed models of site topography for pre-construction engineering studies. The generated data can also be transferred to any CAD or BIM software so that engineers can immediately start working from a 3D model.
- As data collection by drones is easily repeatable at low cost, images can be taken at regular intervals and overlaid on the original blueprints to assess whether the construction work is moving according to specifications.
- High resolution orthophotos enable surveyors to perform highly-accurate distance and surface measurements. With 3D modelling software, it is also possible to obtain volumetric measurements from the very same images. This fast and inexpensive method of volume measurement is particularly useful for calculating stocks in mines and quarries for inventory or monitoring purposes.
- With a drone, surveyors can capture many more data points, hence more accurate volume measurement. They can also do this in a much safer way than if they had to manually capture the data by going-up and down a stockpile. Since drones are capturing the data from above, operations on site won't be interrupted. The short acquisition time enables capturing a snapshot on a specific point in time.
- The development of increasingly dense and complex urban areas requires intensive planning and therefore time-consuming and expensive data collection. Thanks to drones, urban planners can collect large amounts of up-to-date data in a short period of time and with far less staff. The images produced in this way allow planners to examine the existing social and environmental conditions of the sites and consider the impact of different scenarios.

3.1.7 Direct Reading Grade Rods

- Direct Reading Grade Rods or True Elevation Grade Rods (also known as Lenker Rod style) is a Grade Rod with a face that allows the user to set the actual benchmark elevation, or any elevation, above or below the benchmark into the rod. The actual true elevation above sea level is read on these rods, thus eliminating errors caused by adding or subtracting. Since no math is involved, grade shots are much faster than conventional style rods. On most job sites, the bench elevation is typically established by survey, and all elevations on that project are above or below that primary bench elevation.

- A Grade Rod or levelling rod, is a graduated rod used to determine differences in elevation.
- Direct Reading or True Elevation (also known as Lenker style) is a Grade Rod with a moveable tape face allowing you to set the actual benchmark elevation, or any elevation, above or below the bench into the rod.
- Direct Reading or True Elevation (also known as Lenker style) is a Grade Rod with a moveable tape face allowing you to set the actual benchmark elevation, or any elevation, above or below the bench into the rod. The actual true elevation above sea level is read on these rods, thus eliminating errors caused by adding or subtracting.
- Since no math is involved, grade shots are much faster than conventional style rods. On most jobsites, the bench elevation is typically established by survey, and all elevations on that project are above that or below primary bench elevation.

3.1.8 3D Laser Scanning

- 3D Laser Scanning is a non-contact, non-destructive technology that digitally captures the shape of physical objects using a line of laser light. In other words, 3D laser scanning is a way to capture a physical object's exact size and shape into the computer world as a digital 3-dimensional representation.
- 3D scanning is the process of analyzing a real-world object or environment to collect data on its shape and possibly its appearance (for example, colour). The collected data can then be used to construct digital 3D models. Collected 3D data is useful for a wide variety of applications.
- 3D Laser Scanning is an increasingly popular tool for collecting vast amounts of accurate spatial data within a short amount of time. This makes laser scanners a popular piece of equipment within surveying, film and archaeology companies where on site time may be restricted.
- In essence these scanners collect millions of individual point measurements within minutes. The measurements are then plotted within a single XYZ co-ordinate system to form a 'point cloud' of the object's external surface. With the addition of GPS data these points can be geo-referenced and transformed into a global reference system. Multiple data clouds collected from different viewpoints can also be combined ("registered") using common features in order to create one 3D dataset.

  For example, a building could be scanned to include all external walls and then registered together to include scan data from inside the building.
- The resulting 3D data can be used for visualisation, modelling and planning in 3D for example, BIM, geomorphological change and heritage projects. Additionally the point cloud data can be exploited in order to take any number of measurements without having to physically take each measurement on site. Furthermore, point cloud data can form the basis of a rich dataset containing more than purely spatial information (XYZ).

  For example, intensity or photographic imagery could be applied to create a coloured point cloud and 3D vector data can be over laid to map pipes and cables.
- Laser scanning uses controlled deflection of steered laser beams to measure distance between objects. These rapidly captured measurements are used to construct digital three-dimensional (3D) models of objects such as buildings, landscapes and other environments.
- Surveying allows you to determine, with pin point accuracy, the co-ordinates and terrestrial position of objects. 3D laser scanning and surveying projects require specific expertise and knowledge of state-of-the-art equipment as well as dimensional control and land surveying practices.
- High definition 3D digital-capture techniques help minimize operational shutdown and mitigate risks by capturing complete as-built field construction data without causing disruption to operations.

3.1.9 Laser Level

- A laser level is a control tool used in surveying and construction. A laser beam projector can be fixed to a tripod which is then leveled according to the accuracy of the device. The device will project a fixed green or red beam along the horizontal or vertical axis. There a many different types of laser levels to consider when deciding to buy or hire. The most common levels are line and dot lasers, grade lasers and pipe lasers.
- Laser levels can be used for several reasons and should be considered a much-needed instrument for home use as well as in different industries. Uses vary from indoor applications and outdoor applications.
- Laser levels are used daily as the go-to tool for contractors when referencing plumb, level, and square.
- There are different types of construction lasers are dot lasers, line lasers, and rotary lasers.
  - Dot Lasers (Plumb Lasers).
  - Line Laser Levels (Line Generators).
  - Rotary Laser Levels.
- Self-leveling units offer a greater degree of accuracy. They work best when placed on a surface that the user determines is "close to level." You can use a bubble vial to manually level the unit before the unit's self-levelling mechanism takes over. The laser component hangs like a pendulum inside the level.

Some indoor application of Laser Levels include:

- ➢ Levelling floors.
- ➢ Easily check the height of doors and windows.
- ➢ Align shelves, cabinets, trim etc.
- ➢ Measuring distances between objects.

Some outdoor applications of laser levels include:

- ➢ Basic surveying.
- ➢ Site layout.
- ➢ Check the elevation of land.
- ➢ Plumbing.

- Laser levels were designed to speed up the setting up process in the workplace. Using this device is quick, easy and reliable. The accuracy over long distance is far more superior to that of a simple spirit level, which is suitable for shorter distances, but will give less accurate results for longer distance leveling.

Laser level accessories:

- **Tripod :** These are useful tools to have in the laser level world. It can help to provide a level flooring, making results more accurate.
- **Laser Line Detector :** These tools work alongside laser levels and can provide acoustic or optical signals that make it easier to find laser lines.
- **Batteries :** Strong rechargeable batteries are beneficial, especially if laser levels will be going onsite. Some batteries offer up to 2 days continuous operation.

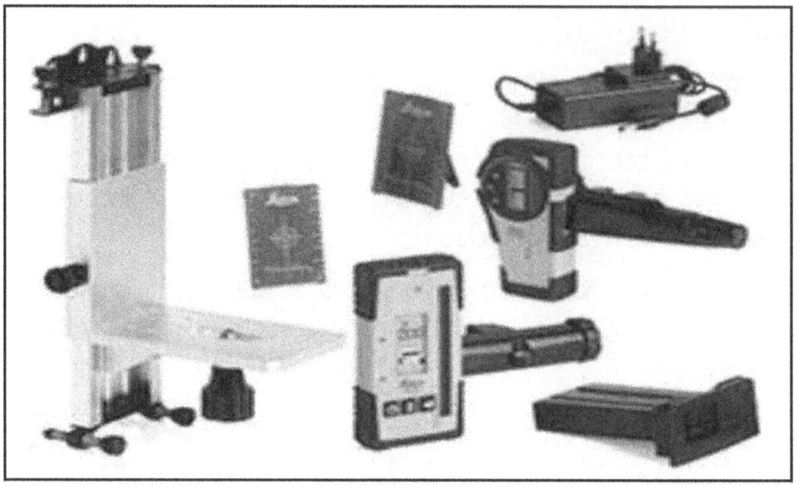

Fig. 3.1

- **Chargers :** Laser level batteries are rechargeable so it is essential to carry chargers to make sure you can use it on site.
- **Remote Controls :** This product will allow you to adjust settings such as rotational speed, auto leveling, manual leveling etc. remotely.
- **Leveling Staff :** Ranging from E-type, bar coded, standard invar and industrial invar, there are many leveling staffs to choose from to assist you with leveling. Our NEDO Fibreglass Leveling Staff is especially sturdy and durable and is ideal for leveling close to power lines and railways.

3.2 CONSTRUCTION EQUIPMENT

- Heavy construction equipment is used for various purposes in large projects. Selection of different types of heavy equipment depends on the size of the work and economy of the project. These make construction process easier and faster.

3.2.1 Earth Moving Equipment

- Earthmoving equipment is heavy equipment, typically heavy-duty vehicles designed for construction operations which involve earthworks. They are used to move large amounts of earth, to dig foundations for landscaping and so on.
- Six types of earth moving equipment on the market.
    1. Excavators : Excavators can be utilised over wheels or treads and are used most often to shovel dirt or lift heavy pieces of machinery.
    2. Backhoe Loaders.
    3. Bulldozers.
    4. Skid Steer Loaders.
    5. Motor Graders.
    6. Trenchers.
- Excavating machine, any machine, usually self-powered, that is used in digging or earth-moving operations of some kind; the power shovel, bulldozer, and grader (q.q.v.) are examples. Excavator.

3.2.1.1 Skid and Crawler Loaders

- Skid-steer loaders are among the most versatile machines available in the construction industry. They are small and nimble and can turn within their own footprint, just like a tank, making them ideal for work in confined spaces or in areas where construction activity has been finalized.

- Skid-steer loaders are wheel-driven and offer good traction in snow and mud. The wheels also minimize soil compaction and damage to finished areas. These handy loaders come in a few different sizes and can be outfitted with a variety of attachments to perform digging, drilling, compacting, log grappling, snow blowing, jack-hammering, and other tasks.
- As the name suggests, these are tyre mounted small-sized machines which can skid on their own axis and therefore, are highly applicable in space-constrained construction sites. The tread pattern used in their tyres ensures that they have minimal impact on finished construction sites and that they limit soil compaction.
- These machines provide a good grip on snow as well as mud. Most skid-steer loaders commonly available in the market can be retrofitted with different attachments and can be used for a wide range of applications such as moving mud or snow, excavation works, compacting soil, drilling holes, lifting loads etc.
- A crawler loader is a cross between an excavator and a backhoe or bucket loader. Its tracks give it excellent stability, and its versatile bucket makes it useful for moving soil and debris and for loading material onto trucks.
- Crawler loaders are also used for excavation on relatively small projects. For larger-scale work, a hydraulic excavator is often used instead of a crawler loader.
- Loaders are used in construction site to load the material onto dumpers, trucks etc. The materials may be excavated soil, demolition waste, raw materials etc. A loader contain large sized bucket at its front with shorter moving arm.
- Loader may be either tracked or wheeled. Wheeled loaders are widely used in sites while tracked or crawled loaders are used in sites where wheeled vehicles cannot reach.

3.2.1.2 Trenchers

- Trenchers are earth moving machines which are used for digging trenches typically to be used for piping, cabling or drainage purposes. The material that has been dug up is deposited besides the trench by the conveyor system of the trencher.
- Different types of trenchers are available which vary according to their applicability, construction requirement or the sturdiness of the surface to be trenched.
- Trenchers generally are mounted either on tracks or on wheels.
- Trenchers or Trenching machines are used to excavate trenches in soil. These trenches are generally used for pipeline laying, cable laying, drainage purposes etc. Trenching machines are available in two types namely chain trenchers and wheeled trenchers.
- Chain trenchers contain a fixed long arm around which digging chain is provided. Wheeled trenchers contain a metal wheel with digging tooth around it. To excavate hard soil layers, wheeled trenchers are more suitable. Both types of trenchers are available in tracked as well as wheeled vehicle forms.
- True to its name, a trencher is used to dig trenches, usually narrow trenches for piping and cabling. Trenchers come in many different types and sizes, from small walk-behind versions to very large trenching machines that can cut into asphalt pavement and other hard surfaces.
- The trencher has a conveyor system that carries the excavated material and deposits it onto the ground next to the trench. Trenchers can use different digging implements depending on the depth of the trench and material being excavated.

3.2.1.3 Scrappers

- A scraper can move dirt and aggregates quickly around a construction site and is used primarily for digging and leveling large areas of land. Scrapers are large heavy equipment designed for open areas, where they can run at high speeds for cut and fill operations. Large self-propelled scrapers are called motor scrapers. There are also smaller scrapers, called pull scrapers, that are pulled behind another vehicle.

- On any construction site you will see a wide variety of heavy equipment, and one of the largest is scrapers. While some machines you see on sites, such as bulldozers and backhoes tell you just what they are designed to do by their appearance, the scraper, with its upfront cab pulling a large wagon, does not give anything away.
- Within that large, heavy wagon section of any scraper is a gated front with a blade at the bottom. With the gate open, the blade scrapes up material as the machine moves forward, forcing the excavated material into the wagon itself, until it is filled. The gate is then closed, and the scraper can transport that material away to be disposed of. If you imagine a carpenter's plane as it shaves off wood, that is the best way to describe how a scraper operates.
- Wheel Tractor Scrapers are earth moving equipment used to provide flatten soil surface through scrapping. Front part contains wheeled tractor vehicle and rear part contain a scrapping arrangement such as horizontal front blade, conveyor belt and soil collecting hopper.
- When the front blade is lowered onto the ground and vehicle is moved, the blade starts digging the soil above the blade level and the soil excavated is collected in hopper through conveyor belt. When the hopper is full, the rear part is raised from the ground and hopper is unloaded at soil dump yard.
- The scraper is a large piece of equipment which is used in mining, construction, agriculture and other earthmoving applications. The rear part has a vertically moveable hopper (also known as the bowl) with a sharp horizontal front edge.
- The hopper can be hydraulically lowered and raised. When the hopper is lowered, the front edge cuts into the soil or clay like a plane and fills the hopper. When the hopper is full it is raised, and closed with a vertical blade (known as the apron). The scraper can transport its load to the fill area where the blade is raised, the back panel of the hopper, or the ejector, is hydraulically pushed forward and the load tumbles out. Then the empty scraper returns to the cut site and repeats the cycle.

Configurations

- Open bowl: usually requires a push-cat (bulldozer or similar) to assist in loading.
- Elevating scraper: self-loading as it uses an elevator to load material; requires no push-cat.
- Tandem scrapers: separate tractor and scraper engines provide better traction in steep or slippery areas; a push cat is required except when loading loose materials.
- Tandem Push-Pull: concentrates the combined horsepower of two such machines onto one cutting edge. The push-pull attachment allows two individual scrapers to act as a self-loading system, typically loading both machines in less than a minute, one after the other.
- Auger: uses vertically mounted auger in the bowl to pull material upwards.
- Pull type scraper: uses agricultural tractor, articulated dump truck, or bulldozer to pull. Pull type scrapers can be utilized individually or two or three units can be pulled behind a single tractor.

3.2.1.4 Wheeled Loading Shovels

- Wheeled loading shovel is a vehicle attached with shovel designed to perform heavy-duty tasks in earthwork and construction operations.
- This shovel has specific capabilities to accomplish the loading tasks. Depending upon various sizes, tasks, and design requirement, this wheeled loader is furnished as skid-steer, payloader, bucket loader, skip loader, and wheel loader.
- It is used to transport materials from place to another. It is equipped with a front mounted bucket, which is supported by a boom structure. For lifting, it loads the material into the bucket by forward motion of the machine. After lifting the material, it transports the materials to its respective destination.
- The growth of the wheeled loading shovels market is driven by innovations and developments in construction equipment market. However, high cost of the wheeled loading shovels machinery as compared to other loading machinery hinders the market growth.

- On the contrary, rapid growth in the construction and mining sector in the Asia-Pacific region, specifically in countries such as China and India, due to favorable government investments and economic growth is expected to offer lucrative opportunities for the market.
- The global wheeled loading shovels market is segmented on the basis of type, net operating power, type of bucket, end user, and region.
- Based on type, the market is segmented into front loader, backhoe loader, compact loader, armored wheel loader, and others (electric rope shovels, skid loaders, track loaders, swing loader, and others).
- On the basis of net operating power, the market is categorized into 2040 hp, 4080 hp, 80120 hp, 120160 hp, and 160 hp and above. By types of bucket, it is classified into wedge bottom bucket, flat bottom bucket, backhoe bucket, demolition bucket, excavating buckets, and others (high dump buckets, multi-purpose buckets, waste handling, woodchip, and others).
- The key players that operating in the market include Caterpillar Inc., Hitachi Ltd., Deere and Company, Kawasaki Heavy Industries, Limited, JCB, Komatsu Ltd., Larsen and Toubro Limited, and New Holland Construction.

Key Benefits for Wheeled Loading Shovels Market:

- The study provides an in-depth analysis of the global wheeled loading shovels industry along with the current and future trends to elucidate imminent investment pockets.
- Information about key drivers, restraints, opportunities, and their impact analysis on the market has been provided.
- Porter's five forces analysis illustrates the potency of buyers and suppliers that operate in the industry.
- The quantitative analysis of the global wheeled loading shovels market from 2017 to 2025 has been provided to determine the market potential.

3.2.2 Advanced Plastering Machine

- General description of the project is given as, "Quick plastering of walls automatically by pumping the cement mix from funnel and plaster it on wall in vertical and horizontal movement using microcontrollers by controlling the stepper motor. It is cost effective and reduces human effort".
- The cement mix which is poured in a funnel is pumped using centrifugal pump to a hopper.
- Inside the hopper, there will be a gear pump which pumps the cement mix as an output to wall through the cylinder slit.
- The cement mix which came out will be captured by the metal plate and it forcibly sticks to the wall.
- The cement mix will be punched by metal plate using rollers.
- The stuck cement mix will also be smudged using metal plate and rollers.
- To make this as automation, we interface the sensors to detect cement flow, the stepper motor for the movement of the header unit in rail guides and the AC motor to control the flow with micro-controllers.
- Here we can give the vertical and horizontal distance as the input for the movement of header unit.
- Advanced plastering machine innovative machine is unique and perhaps one kind of automated plastering machinery ideally suitable for the construction/building industry.
- It works with conventional cement mortar which brings it to a smooth, flat finish with variable and adjustable thickness to suit each application.
- It can plaster the wall automatically by moving-up and down in vertical direction.
- It can be plastered by one-time in vertical direction. The thickness of the ash/gypsum salleri can be adjusted.
- It has special design for adjusting the thickness of plastering/salleri/cement mix.

- It has two rails for rising and moving automatically, therefore it can be used for different height and width of the wall.
- It has large capacity hopper and you can put the ash/lime/gypsum in it one-time.
- It has micro-controllers for controlling the motor to automatic extend of cement flow and to automatic movement of hopper for cement discharge/plastering.
- It is easy to operate. One or two person can operate.
- Easy to move, without removing any parts of the machine and there are wheels under the machine for easy movement.

Applications:

- Automated plastering machine is unique and perhaps one kind of automated plastering machinery ideally suitable for the construction/building industry.
- Automated plastering machine works with conventional cement mortar which brings it to a smooth, flat finish with variable and adjustable thickness to suit each application.
- Automated plastering machine makes rendering easier, faster, and effortless as compare to manual application.
- This idea can also further enhanced by interfacing LCD display and Keypad for making the process without external source.
- This time and money saving machine, keeps up with the ever changing world of building automation.

3.2.3 Bridge Launcher

- The bridge launcher is the equipment that places the preformed beam into the prefabricated piers. The bridge machine belongs to the crane category, because its main function is to lift the beam, and then transported to the position and then put down. But it is very different from the general sense of the crane. The requirements of the harsh conditions, and the existence of beam on the line, or called the vertical shift. Bridge machine is divided into erecting Highway Bridge, conventional railway bridge, passenger railway bridge, and several.

Application range of Bridge Launcher

- The bridge launcher is designed for highway and railway. According to the situation of present railway with less sub grade, more bridges and tunnels which connected too closely, our company design and manufacture the new box-girder bridge launcher with double girder.
- It is mainly used for hoisting (lifting) of concrete girder in bridge construction (bridge building) of general railway and highway.

Main features and advantages of Bridge Launcher

- According to users' demands, the new type Bridge Girder Erection Machine which integrates carrying with erecting is designed.

Features

- The machine can accomplish the girder's carrying with erecting at the same time.
- The machine can erect 4 pieces of girders with another two cranes one day, saving construction cycle.

Another advantage:

- If the concrete girder is between the two piers, just exchange the front and rear legs, it can continue to work, no extra assembly, shortening construction period.

3.3 MATERIAL HANDLING EQUIPMENT

- Material handling equipment is mechanical equipment used for the movement, storage, control and protection of materials, goods and products throughout the process of manufacturing, distribution, consumption and disposal.
- The different types of handling equipment can be classified into four major categories: transport equipment, positioning equipment, unit load formation equipment, and storage equipment.

3.3.1 Cranes

- A wide range of different types of crane can be used on construction projects as a means for lifting and moving materials and other building elements from place-to-place. Cranes can range from a simple rope pulley to a tower crane fixed to the top of a skyscraper.
- A launching gantry is a special-purpose mobile gantry crane used in bridge construction. It is used to install precast box girders in highway and high-speed rail bridge. A launching gantry (described also by other terms including beam launcher, girder launcher, bridge building crane, and bridge-building machine,
- A crane is a type of machine, generally equipped with a hoist rope, wire ropes or chains, and sheaves, that can be used both to lift and lower materials and to move them horizontally. It is mainly used for lifting heavy things and transporting them to other places.

The choice of crane for a particular site will depend a number of different factors, including:

- The type of loads that need to be lifted.
- The height and horizontal distances that need to be covered.
- The radius of the swing.
- The time period of the lifting operations.
- The utilization factor (the ratio of time spent in use compared to the total time it could be in use).
- The degree of mobility required.

  The type of plant that will be used in conjunction with the crane, such as concreting plant.

Cranes be classified as:

- Mobile.
- Static.
- Tower.

Mobile cranes

- Mobile cranes can be moved within a site, giving them great flexibility. They can be either mobile wheeled, truck-mounted, track-mounted or gantry-based.

Mobile wheeled cranes

- Mobile wheeled cranes are mounted on a wheeled chassis with stabilisers that can be used when lifting to prevent movement. Generally, crane is controlled and driven by an operator inside the cab. The slewing ring beneath the cab allows for a 360-degree turning circle, and the boom extends upwards supported by suspension ropes.
- Lifting capacity: Tends to vary from 3 - 50 tonnes, but is generally around 10 tonnes.
- Speed limits apply depending on the type of vehicle and road.
- Mobile wheeled cranes are commonly used for moving moderately heavy loads, equipment and other plant in goods yards or storage areas.

Truck-mounted cranes

- Truck-mounted cranes are mounted on a truck or lorry specially adapted to carry an increased load. The lorry can be driven from a front cab as a conventional vehicle, as well as having additional controls for a lattice mast or telescopic boom which extends in sections. Fly jib attachments can be used to increase lifting height.
- Lifting capacity: Can vary from 5 - 2,000 tonnes 35 - 100 being most commonly used.
- Speed limits apply depending on the type of vehicle and road.
- As this type of crane is very transportable and has a short site preparation time they are commonly used for short hire periods.

- Track-mounted cranes are mounted on a diesel powered crawler unit together with a lattice mast that can accommodate a fly jib attachment with additional lifting ropes to allow for better site coverage. They can move at low speed around sites but need to be transported to and rigged on the site.
- Lifting capacity: Tends to vary from 5 – 30 tonnes.
- Average speed: Usually does not exceed 2 km/h.
- The flexibility of this crane is its main advantage, as it can be adapted to act as a dragline with the addition of a winch drum at the front. The tracks allow for it to be operational on poor ground conditions.

Gantry crane

- Also known as portal cranes, gantry cranes typically consist of two 'A' frames connected by a lattice cross member which straddles the work area. The lifting gear is suspended from the horizontal cross member and can move along it on rails. On small gantry cranes, the 'A' frame is wheel mounted, whereas larger cranes are mounted on powered bogies that run on rail tracks. Larger cranes tend to have the lifting gear mounted with a driving cab on the cross member.
- Lifting capacity: Small gantry cranes have a 10 tonne capacity, but larger versions can lift-up to 100 tonnes.
- This form of crane is commonly used for repetitive work on low to medium-rise developments, or in stock yards to move equipment and materials.

Static cranes

- Despite having a similar appearance to the traditional tower cranes, static cranes differ in that the lattice mast is mounted on the slewing ring meaning that the whole tower rotates from one fixed position.
- They are fitted with a luffing jib which can be raised through 90-degrees from horizontal, with a trolley and hoist block that can move along the length of the jib.
- Lifting capacity: These will vary depending on the length of the jib, the position to which the jib is raised, and the position of the hoist block along the length.
- Static cranes tend to be used for low capacity lifting and are beneficial in that they tend to allow for a closer approach to the face of a building.

Tower cranes

- Tower cranes are a familiar sight on most construction projects. Tower crane operators are capable of seeing most of the lifting operations from the cab, although a banksman is required at ground level both for overseeing the loading of the crane and for issuing signals and guidance to the operator. They are usually assembled and erected on-site with a horizontal or luffing jib.

Common features of tower cranes include:

- Lattice sectionalised tower.
- Jib (typically 25 m long) fitted with precast concrete counterweight blocks.
- Access ladder housed within the tower or attached to an outside face.
- Stay rods and support frame.
- Cab at the top of the tower in which an operator sits.
- Slewing ring which allows for rotation of the jib.
- Trolley and hoist block which can move along the length of the jib.

Tower cranes come a number of different forms:

- Self-supporting static.
- Supporting static.
- Travelling or rail-mounted.
- Climbing.

3.3.2 Conveyors

- A conveyor system is a common piece of mechanical handling equipment that moves materials from one location to another. Conveyors are especially useful in applications involving the transportation of heavy or bulky materials.
- As far as growth is concerned the material handling and conveyor system makers are getting utmost exposure in the industries like automotive, pharmaceutical, packaging and different production plants. The portable conveyors are likewise growing fast in the construction sector.
- Belt conveyors are generally fairly similar in construction consisting of a metal frame with rollers at either end of a flat metal bed. Rubber conveyor belts are commonly used to convey items with irregular bottom surfaces, small items that would fall in between rollers (for example, a sushi conveyor bar), or bags of product that would sag between rollers.
- The belt is looped around each of the rollers and when one of the rollers is powered (by an electrical motor) the belting slides across the solid metal frame bed, moving the product. In heavy use applications the beds which the belting is pulled over are replaced with rollers.
- The rollers allow weight to be conveyed as they reduce the amount of friction generated from the heavier loading on the belting. The exception to the standard belt conveyor construction is the Sandwich Belt conveyor.
- The Sandwich Belt conveyor uses two conveyor belts, instead of one. These two conventional conveyor belts are positioned face to face, to firmly contain the items being carried in a "sandwich-like" hold.
- Conveyor systems allow quick and efficient transportation for a wide variety of materials, which make them very popular in the material handling and packaging industries.
- They also have popular consumer applications, as they are often found in supermarkets and airports, constituting the final leg of item/bag delivery to customers.
- Many kinds of conveying systems are available and are used according to the various needs of different industries. There are chain conveyors (floor and overhead) as well. Chain conveyors consist of enclosed tracks, I-Beam, towline, power and free, and hand pushed trolleys.
- Conveyor systems are used widespread across a range of industries due to the numerous benefits they provide.
- Conveyors are able to safely transport materials from one level to another, which when done by human labour would be strenuous and expensive.
- They can be installed almost anywhere, and are much safer than using a forklift or other machine to move materials.
- They can move loads of all shapes, sizes and weights. Also, many have advanced safety features that help prevent accidents.
- There are a variety of options available for running conveying systems, including the hydraulic, mechanical and fully automated systems, which are equipped to fit individual needs.
- Conveyor systems are commonly used in many industries, including the Mining, automotive, agricultural, aerospace, electronics, pharmaceutical, chemical, bottling and canning, print finishing and packaging.
- Although a wide variety of materials can be conveyed, some of the most common include food items such as beans and nuts, bottles and cans, automotive components, scrap metal, pills and powders, wood and furniture and grain and animal feed.
- Many factors are important in the accurate selection of a conveyor system. It is important to know how the conveyor system will be used beforehand. Some individual areas that are helpful to consider are the required conveyor operations, such as transportation, accumulation and sorting, the material sizes, weights and shapes and where the loading and pickup points need to be.
- Conveyors are used when material is to be moved frequently between specific points over a fixed path and when there is a sufficient flow volume to justify the fixed conveyor investment.

- Different types of conveyors can be characterized by the type of product being handled: unit load or bulk load; the conveyor's location: in-floor, on-floor, or overhead, and whether or not loads can accumulate on the conveyor.
- Accumulation allows intermittent movement of each unit of material transported along the conveyor, while all units move simultaneously on conveyors without accumulation capability.
- For example, while both the roller and flat-belt are unit-load on-floor conveyors, the roller provides accumulation capability while the flat-belt does not; similarly, both the power-and-free and trolley are unit-load overhead conveyors, with the power-and-free designed to include an extra track in order to provide the accumulation capability lacking in the trolley conveyor. Examples of bulk-handling conveyors include the magnetic-belt, troughed-belt, bucket, and screw conveyors.
- A sortation conveyor system is used for merging, identifying, inducting, and separating products to be conveyed to specific destinations, and typically consists of flat-belt, roller, and chute conveyor segments together with various moveable arms and/or pop-up wheels and chains that deflect, push, or pull products to different destinations.

3.3.3 Hoists

- A hoist is a device used for lifting or lowering a load by means of a drum or lift-wheel around which rope or chain wraps. It may be manually operated, electrically or pneumatically driven and may use chain, fiber or wire rope as its lifting medium. The most familiar form is an elevator, the car of which is raised and lowered by a hoist mechanism. Most hoists couple to their loads using a lifting hook.
- A hoist is also known as a Man-Lift, Buckhoist, temporary elevator, builder hoist, passenger hoist or construction elevator, this type of hoist is commonly used on large scale construction projects, such as high-rise buildings or major hospitals. There are many other uses for the construction elevator.
- Many other industries use the buckhoist for full-time operations, the purpose being to carry personnel, materials, and equipment quickly between the ground and higher floors, or between floors in the middle of a structure. There are three types: Utility to move material, personnel to move personnel, and dual-rated, which can do both.
- The construction hoist is made-up of either one or two cars (cages) which travel vertically along stacked mast tower sections. The mast sections are attached to the structure or building every 25 feet (7.62 m) for added stability. For precisely controlled travel along the mast sections, modern construction hoists use a motorized rack-and-pinion system that climbs the mast sections at various speeds.
- Hoists are used on construction sites to vertically transport materials and/or passengers. They are most commonly powered by diesel engines or electric motors that rotate a drum around which a wire rope is wound. Hoists may also be hydraulically powered, and may use chains as the lifting mechanism rather than wire rope.
- The landing area must be fitted with some form of guard, usually sliding gates, and the base of the hoist, where the winch is situated, should be well guarded to prevent injury.
- Hoists differ from cranes in that cranes move items vertically and horizontally. Hoists differ from lifts in that they are generally used for industrial purposes and are not publicly accessible. Winches are typically used to pull or drag items along level surfaces, rather than lifting them vertically.

Mobile hoists

- Mobile hoists are commonly found on construction sites, and are capable of lifting material loads to heights of up to 30 m. They are designed to be dismantled, folded onto the chassis and moved to another location with relative ease, either under their own power or towed by a haulage vehicle.

- The mast and winch unit is mounted on a platform, typically with a load capacity of 500 kg. This is then stabilised using jacks or outriggers. Extending upwards, a lattice hoist mast is constructed to which sections can be added depending on the height required, together with tie supports fixed to scaffolding or the building frame. A protective screen is placed around the hoist mast, fitted with gates at least 2 m high at all landing levels.

Passenger hoists

- These are designed to lift passengers, although they can also be used for materials as long as the weight is kept within the loading capacity. The type of hoist can vary from a single cage with rope suspension to twin cages with rack and pinion operation mounted on two sides of a static tower. They are usually controlled from within the cage, and there must be additional safety devices to prevent over-run or free-fall.
- A typical passenger hoist cage is 2.7 m high and capable of carrying 12 passengers at a total weight of 1,000 kg. Typical speeds are 40-100 m/min. The hoist tower is generally assembled from 1.5 m-long sections and tied at 12 m centres to the face of the structure.

Small Package Hoists

- Small hoists consisting of a small lifting arm, simple electric motor and wire rope can be bolted to a structure or to scaffolding and used for light-lifting operations of around 500 kg.

3.3.4 Forklifts

- A forklift is a powered industrial truck used to lift and move materials over short distances. The forklift was developed in the early 20th century by various companies.
- Forklifts have become an indispensable piece of equipment in manufacturing and warehousing.
- Forklifts are one of the most powerful vehicles for builders. Whether you need to move heavy loads in a warehouse or construction field, forklifts can safely lift objects humans cannot, making them indispensable on any job site.
- Since there is a wide selection of styles and models, it can be difficult to select a forklift. Whether you are looking to rent a forklift for your next project or purchase one, our guide covers the different forklift types, the benefits of each equipment, and popular forklift models to help choose the right forklift for your needs.
- Forklifts are incredibly convenient machines. They are prominently used in material handling and construction. They are able to lift incredibly large loads and reach heights that could frighten most people. Because of this, they are a staple of warehouses and they fill the shelves daily.

The most popular forklift types on the market today are:

- Warehouse Forklift.
- Side Loader.
- Counterbalance Forklift.
- Telehandler.
- Industrial Forklift.
- Rough Terrain Forklift.
- Pallet Jack.
- Walkie Stacker.
- Order Picker.
- Reach Fork Truck.

The selection of the appropriate types of forklift

The common features of forklift trucks include their engine, the driver's cab, and the lifting forks. The selection of the appropriate type of forklift will depend on considerations such as:

- The load and lifting height required.
- The use of attachments (such as fork extensions, booms, rotating heads, drum clamps, paper roll clamps, bale clamps, load stabilizers and so on).

- Available power sources (electric batteries, liquefied petroleum gas (LPG), petrol or diesel).
- The terrain.
- The need to protect the operator.
- Exposure to the weather.
- The need for lighting.
- Noise restrictions.

Forklift control and capabilities

- Forklift hydraulics are controlled either with levers directly manipulating the hydraulic valves or by electrically controlled actuators, using smaller "finger" levers for control. The latter allows forklift designers more freedom in ergonomic design.
- Forklift trucks are available in many variations and load capacities. In a typical warehouse setting, most forklifts have load capacities between one and five tons. Larger machines, up to 50 tons lift capacity, are used for lifting heavier loads, including loaded shipping containers.
- In addition to a control to raise and lower the forks (also known as blades or tines), the operator can tilt the mast to compensate for a load's tendency to angle the blades toward the ground and risk slipping-off the forks. Tilt also provides a limited ability to operate on non-level ground. Skilled forklift operators annually compete in obstacle and timed challenges at regional forklift rodeos.

3.3.5 Mobile Concrete Mixer

- A concrete mixer is a device that homogeneously combines cement, aggregate such as sand or gravel, and water to form concrete. A typical concrete mixer uses a revolving drum to mix the components. For smaller volume works, portable concrete mixers are often used so that the concrete can be made at the construction site, giving the workers ample time to use the concrete before it hardens. An alternative to a machine is mixing concrete by hand. This is usually done in a wheelbarrow; however, several companies have recently begun to sell modified tarps for this purpose.
- Mobile concrete mixer is with tyres, so it is convenient to move from site to site, and it is usually used to mix plastic and semi-harsh concrete. They are more suitable for construction site, road, bridge and water-power engineering. Compared to the stationary concrete mixers, it is more convenient to change the installation site according to the need of construction project, particularly small and medium-sized engineering.

Working principle of mobile concrete mixer machine

- Along the wall of the cylinder, installs several mixing blades. When the mixer begins to work, the cylinder spins on its own, uses the mixing blades to break-up, lift, scatter and shock the materials, and relocates all materials and mixing evenly.

Advantages of mobile concrete mixer

1. Small volume and light weight.
2. Convenient movement and long service life.
3. Reasonable and compact structure and attractive appearance.
4. Easy operation and reliable transmission.
5. Smooth function and low energy consumption.
6. Good mixing quality and fast mixing speed.

- There are many different types of mobile concrete mixers. Such as diesel concrete mixer, electric concrete mixer and self-loading concrete mixer.

3.3.6 Paver

- Paver or Asphalt paver is pavement laying equipment which is used in road construction. Paver contains a feeding bucket in which asphalt is continuously loaded by the dump truck and paver distributes the asphalt evenly on the road surface with slight compaction. However a roller is required after laying asphalt layer for perfect compaction.

- A paver (paver finisher, asphalt finisher, paving machine) is a piece of construction equipment used to lay asphalt on roads, bridges, parking lots and other such places. It lays the asphalt flat and provides minor compaction before it is compacted by a roller.

Operation

- The asphalt is added from a dump truck or a material transfer unit into the paver's hopper. The conveyor then carries the asphalt from the hopper to the auger. The auger places a stockpile of material in front of the screed. The screed takes the stockpile of material and spreads it over the width of the road and provides initial compaction.
- The paver should provide a smooth uniform surface behind the screed. In order to provide a smooth surface a free floating screed is used. It is towed at the end of a long arm which reduces the base topology effect on the final surface. The height of the screed is controlled by a number of factors including the attack angle of the screed, weight and vibration of the screed, the material head and the towing force.
- Additional controls are used to correct the slope, crown or super elevation of the finished pavement.
- In order to provide a smooth surface the paver should proceed at a constant speed and have a consistent stockpile of material in front of the screed. Increase in material stockpile or paver speed will cause the screed to rise resulting in more asphalt being placed therefore a thicker mat of asphalt and an uneven final surface. Alternatively a decrease in material or a drop in speed will cause the screed to fall and the mat to be thinner.
- The need for constant speed and material supply is one of the reasons for using a material transfer unit in combination with a paver. A material transfer unit allows for constant material feed to the paver without contact, providing a better end surface. When a dump truck is used to fill the hopper of the paver, it can make contact with the paver or cause it to change speed and affect the screed height.

Concrete Pavers

- Concrete pavers encompass a range of types, including form-riding paving machines, curb-and-gutter machines, and slip form machines.
- The most widely used machine for paving concrete roadways is the mainline slip form paver, which extrudes a paved surface without the use of forms.
- Curb-and-gutter pavers can use variously shaped molds to place other than flat surfaces, such as curbs, roadway median barriers, and bridge parapets.
- Many mainline pavers automatically change paving widths without stopping the paving process.

3.3.7 Road Header

- Road header is a piece of excavating equipment consisting of a boom-mounted cutting head, a loading device usually involving a conveyor, and a crawler travelling track to move the entire machine forward into the rock face.
- A road header machine called a boom type road header. Road headers are advanced; self-forced and extremely powerful rock cutting machines are designed to excavate roadways, tunnels and compartments continuously without using volatile. Powered electro hydraulically, they produce no fumes and are used extensively in mineral mining, coal mining and underground construction projects, where their capacity to excavate the desired profile without causing damaging vibrations is highly valued for both safety environmental reasons.
- The cutting head can be a general purpose rotating drum mounted in line or perpendicular to the boom, or can be special function heads such as jack-hammer like spikes, compression fracture micro-wheel heads like those on larger tunnel boring machines, a slicer head like a gigantic chain saw for dicing up rock, or simple jaw-like buckets of traditional excavators.

- It is a piece of excavating equipment consisting of a boom mounted cutting head, a loading machine usually involving a conveyor and a crawler travelling track to move the whole machine forward into the rock face. The road header also includes separate part, multi-purpose hydraulic cutting heads for increasing on excavators.
- The cutting head can be a general purpose rotating drum mounted in line or perpendicular to the boom or can be particular function heads such as jack hammer like spikes, compression fracture micro wheel heads like those on larger tunnel boring machines like a gigantic chain saw for dicing-up rock or simple jaw like buckets of traditional excavators.
- **Construction road headers:** Road headers for construction are set with powerful, transverse cutter heads proven to give the best cutting performance in a wide range of rock formations. Subway tunnels, shaft sinking, rehabilitation of existing tunnels, road tunnels and excavation of underground caverns are some examples where these machines have established their great flexibility.
- **Mining road headers:** Road headers for mining are prepared with powerful, transverse cutter heads proven to give an unparalleled cutting performance in a wide range formation of rock. Mounted on an extremely robust hydraulic boom they are reliable, rugged, highly creative result for both development and direct production duties.
- **Hydraulic cutting heads:** Hydraulic cutting heads are designed to be increase on suitable excavators quickly and easily. With transverse cutting structures, tried, geometrically optimized and tested on some of the world's most rocky mining and tunneling machines, they are well organized, multi-use tools for various applications including scaling, production, tunnel treatment, roadway leveling, trenching and demolition.

3.3.8 Tunnel Boring Machine

- A **tunnel boring machine (TBM)** is a machine used to excavate tunnels with a circular cross section through a variety of soil and rock strata. They can bore through hard rock, sand, and almost anything in between. Tunnel diameters can range from a metre (done with micro-TBMs) to almost 16 metres to date. Tunnels of less than a metre or so in diameter are typically done using trenchless construction methods or horizontal directional drilling rather than TBM's.
- Tunnel boring machines are used as an alternative to drilling and blasting (D and B) methods in rock and conventional 'hand mining' in soil. TBMs have the advantages of limiting the disturbance to the surrounding ground and producing a smooth tunnel wall. This significantly reduces the cost of lining the tunnel, and makes them suitable to use in heavily urbanized areas. The major disadvantage is the upfront cost. TBMs are expensive to construct, and can be difficult to transport. However, as modern tunnels become longer, the cost of tunnel boring machines versus drill and blast is actually less – this is because tunnelling with TBM's is much more efficient and results in a shorter project.
- The largest diameter TBM, at 15.43 m, was built by Herrenknecht AG for a recent project in Shanghai, China. The machine was built to bore through soft ground including sand and clay. The largest diameter hard rock TBM, at 14.4 m, was manufactured by The Robbins Company for Canada's Niagara Tunnel Project. The machine is currently boring a hydroelectric tunnel beneath Niagara Falls, the machine has been named "Big Becky" in reference to the Sir Adam Beck hydroelectric dams to which it is tunneling to provide an additional hydroelectric tunnel.

Hard rock TBMs

- In hard rock, either shielded or open-type TBMs can be used. All types of hard rock TBM's excavate rock using disc cutters mounted in the cutter head. The disc cutters create compressive stress fractures in the rock, causing it to chip away from the rock in front of the machine, called the tunnel face. The excavated rock, known as muck, is transferred through openings in the cutter head to a belt conveyor, where it runs through the machine to a system of conveyors or muck cars for removal from the tunnel.

- Open-type TBMs have no shield, leaving the area behind the cutter head open for rock support. To advance, the machine uses a gripper system that pushes against the side walls of the tunnel. The machine can be continuously steered while gripper shoes push on the side-walls to react the machine's forward thrust. At the end of a stroke, the rear legs of the machine are lowered, the grippers and propel cylinders are retracted. The retraction of the propel cylinders repositions the gripper assembly for the next boring cycle. The grippers are extended, the rear legs lifted, and boring begins again. The open-type, or Main Beam, TBM does not install concrete segments behind it as other machines do. Instead, the rock is held-up using ground support methods such as ring beams, rock bolts, shotcrete, steel straps, and wire mesh (Stack, 1995).
- In fractured rock, shielded hard rock TBM's can be used, which erect concrete segments to support unstable tunnel walls behind the machine. Double Shield TBMs are so called because they have two modes; in stable ground they can grip against the tunnel walls to advance forward. In unstable, fractured ground, the thrust is shifted to thrust cylinders that push-off against the tunnel segments behind the machine. This keeps the significant thrust forces from impacting fragile tunnel walls. Single Shield TBM's operate in the same way, but are used only in fractured ground, as they can only push off against the concrete segments (Stack, 1995).

Soft ground TBMs

- In soft ground, there are two main types of TBM's: Earth Pressure Balance Machines (EPB) and Slurry Shield (SS). Both types of machines operate like Single Shield TBM's, using thrust cylinders to advance forward by pushing off against concrete segments. Earth Pressure Balance Machines are used in soft ground with less than 7 bar of pressure. The cutter head does not use disc cutters only, but instead a combination of tungsten carbide cutting bits, carbide disc cutters, and/or hard rock disc cutters. The EPB gets its name because it is capable of holding-up soft ground by maintaining a balance between earth and pressure. The TBM operator and automated systems keep the rate of soil removal equal to the rate of machine advance. Thus, a stable environment is maintained. In addition, additives such as bentonite, polymers and foam are injected into the ground to further stabilize it.
- In soft ground with very high water pressure and large amounts of ground water, Slurry Shield TBMs are needed. These machines offer a completely enclosed working environment. Soils are mixed with bentonite slurry, which must be removed from the tunnel through a system of slurry tubes that exit the tunnel. Large slurry separation plants are needed on the surface for this process, which separate the dirt from the slurry so it can be recycled back into the tunnel.
- While the use of TBMs relieves the need for large numbers of workers at high pressures, a caisson system is sometimes formed at the cutting head for slurry shield TBM's. Workers entering this space for inspection, maintenance and repair need to be medically cleared as "fit to dive" and trained in the operation of the locks.

Back-up systems

- Behind all types of tunnel boring machines, inside the finished part of the tunnel, are trailing support decks known as the back-up system. Support mechanisms located on the back-up can include: conveyors or other systems for muck removal, slurry pipelines if applicable, control rooms, electrical systems, dust removal, ventilation and mechanisms for transport of pre-cast segments.

Urban tunnelling and near surface tunneling

- Urban tunnelling has the special challenge of requiring that the ground surface be undisturbed. This means that ground subsidence must be avoided. The normal method of doing this in soft ground is to maintain the soil pressures during and after the tunnel construction. There is some difficulty in doing this, particularly in varied strata (for example, boring through a region where the upper portion of the tunnel face is wet sand and the lower portion is hard rock).
- TBMs with positive face control, such as EPB and SS, are used in such situations. Both types (EPB and SS) are capable of reducing the risk of surface subsidence and voids if operated properly and if the ground conditions are well documented.

- When tunnelling in urban environments, other tunnels, existing utility lines and deep foundations need to be addressed in the early planning stages. The project must accommodate measures to mitigate any detrimental effects to other infrastructure.

1- Slurry Pressure Balance (SPB) TBM

- The basic principle of this TBM is to maintain the face pressure during the excavation phase by filling the working chamber, located behind the cutter head, with slurry.

Advantages

- Allows soft, wet, or unstable ground to be tunnelled with a speed and safety not previously possible
- Suitable for ground with high water pressures (below water table)
- Limits ground settlement and produces a smooth tunnel wall. This significantly reduces the cost of lining the tunnel, and makes it suitable to use in heavily urbanized areas.

Disadvantages

- The major disadvantage is the upfront capital cost. TBM's are expensive to construct, difficult to transport, require significant backup systems and power.
- Drive can be hindered by large stones and boulders

2- Earth Pressure Balance (EPB) TBM

- This is a mechanised tunnelling method in which spoil is admitted into the tunnel boring machine (TBM) via a screw conveyor arrangement which allows the pressure at the face of the TBM to remain balanced without the use of slurry.

Advantages

- Allows soft, wet, or unstable ground to be tunnelled with a speed and safety not previously possible
- Limits ground settlement and produces a smooth tunnel wall. This significantly reduces the cost of lining the tunnel, and makes it suitable to use in heavily urbanized areas

Disadvantages

- The major disadvantage is the upfront capital cost. TBMs are expensive to construct, difficult to transport, require significant backup systems and power.

3-Hard Rock TBM

- This method involves the use of a Tunnelling machine with a shield and cutter head suitable for hard rock.

Advantages

- They offer a continuous and controlled means of tunnelling capable of high rates of advance under favourable conditions.

Disadvantages

- The major disadvantage is the upfront capital cost. TBMs are expensive to construct, difficult to transport, require significant backup systems and power.
- Their applicability is limited to long tunnels where the high rates of advance and tunnel quality can offset their high capital cost.

Important Points

- LiDAR, which stands for Light Detection and Ranging, is a remote sensing method that uses light in the form of a pulsed laser to measure ranges (variable distances) to the Earth.
- As it turns out, the accuracy of the laser ranging is actually one of the most accurate components in the whole system. Although airborne LiDAR systems may not be this accurate, as it turns out the error contribution from the laser ranging is by far one of the least significant sources of overall error.

- LiDAR is a surveying method that measures distance to a target by illuminating the target with laser light and measuring the reflected light with a sensor. Differences in laser return times and wavelengths can then be used to make digital 3-D representations of the target.
- G.P.S., which stands for Global Positioning System, is a radio navigation system that allows land, sea, and airborne users to determine their exact location, velocity, and time 24 hours a day, in all weather conditions, anywhere in the world.
- G.P.S. systems are extremely versatile and can be found in almost any industry sector. They can be used to map forests, help farmers harvest their fields, and navigate airplanes on the ground or in the air. G.P.S. systems are used in military applications and by emergency crews to locate people in need of assistance.
- A Geographic Information System is a system designed to capture, store, manipulate, analyze, manage, and present spatial or geographic data. G.I.S. applications are tools that allow users to create interactive queries, analyze spatial information, edit data in maps, and present the results of all these operations.
- The G.I.S. is used to study geologic features, analyze soils and strata, assess seismic information, and or create three dimensional (3D) displays of geographic features. GIS can be also used to analyze rock information characteristics and identifying the best dam site location.
- Photogrammetry is the science of making measurements from photographs. The input to photogrammetry is photographs, and the output is typically a map, a drawing, a measurement, or a 3D model of some real-world object or scene.
- There are two types of Photogrammetry, Aerial Photogrammetry and Terrestrial (Close Range) Photogrammetry. Aerial digital photogrammetry, often used in topographical mapping, begins with digital photographs or video taken from a camera mounted on the bottom of an airplane.
- Photogrammetry is the use of photography in surveying and mapping. It's used to measure distances between objects, and photogrammetry software can utilize images to create 3D renderings. While in its early days it was used mostly for topographic mapping, it's recently been used in a number of unexpected ways.
- A typical unmanned aircraft is made of light composite materials to reduce weight and increase maneuverability. Drones are controlled by remote ground control systems (GSC) and also referred to as a ground cockpit. An unmanned aerial vehicle system has two parts, the drone itself and the control system.
- A Grade Rod or levelling rod, is a graduated rod used to determine differences in elevation.
- Direct Reading or True Elevation (also known as Lenker style) is a Grade Rod with a moveable tape face allowing you to set the actual benchmark elevation, or any elevation, above or below the bench into the rod.
- 3D Laser Scanning is a non-contact, non-destructive technology that digitally captures the shape of physical objects using a line of laser light. In other words, 3D laser scanning is a way to capture a physical object's exact size and shape into the computer world as a digital 3-dimensional representation.
- 3D scanning is the process of analyzing a real-world object or environment to collect data on its shape and possibly its appearance (for example, colour). The collected data can then be used to construct digital 3D models. Collected 3D data is useful for a wide variety of applications.
- Laser levels are used daily as the go-to tool for contractors when referencing plumb, level, and square.
- There are different types of construction lasers are dot lasers, line lasers, and rotary lasers.
  - Dot Lasers (Plumb Lasers).
  - Line Laser Levels (Line Generators).
  - Rotary Laser Levels.

- Self-levelling units offer a greater degree of accuracy. They work best when placed on a surface that the user determines is "close to level." You can use a bubble vial to manually level the unit before the unit's self-levelling mechanism takes over. The laser component hangs like a pendulum inside the level.
- Earthmoving equipment is heavy equipment, typically heavy-duty vehicles designed for construction operations which involve earthworks. They are used to move large amounts of earth, to dig foundations for landscaping and so on.
- Six types of earth moving equipment on the market:
    1. Excavators. Excavators can be utilised over wheels or treads and are used most often to shovel dirt or lift heavy pieces of machinery.
    2. Backhoe Loaders.
    3. Bulldozers.
    4. Skid Steer Loaders.
    5. Motor Graders.
    6. Trenchers.
- Excavating machine, any machine, usually self-powered, that is used in digging or earth-moving operations of some kind; the power shovel, bulldozer, and grader (qq.v.) are examples. Excavator.
- A launching gantry is a special-purpose mobile gantry crane used in bridge construction. It is used to install precast box girders in highway and high-speed rail bridge. A launching gantry (described also by other terms including beam launcher, girder launcher, bridge building crane, and bridge-building machine,
- A crane is a type of machine, generally equipped with a hoist rope, wire ropes or chains, and sheaves, that can be used both to lift and lower materials and to move them horizontally. It is mainly used for lifting heavy things and transporting them to other places.
- A conveyor system is a common piece of mechanical handling equipment that moves materials from one location to another. Conveyors are especially useful in applications involving the transportation of heavy or bulky materials.
- A tunnel boring machine (TBM) is a machine used to excavate tunnels with a circular cross section through a variety of soil and rock strata. They may also be used for microtunneling.
- With its rotating cutting wheel, the tunneling machine breaks the material from the tunnel phase. The material is then transferred to the belt conveyer system in the rear of the shield via a screw conveyer while the hydraulic cylinders press the machine forward continuously.
- A road header, also called a boom-type road header, road header machine, road header or just header machine, is a piece of excavating equipment consisting of a boom-mounted cutting head, a loading device usually involving a conveyor, and a crawler travelling track to move the entire machine forward into the rock face.
- The cutting head can be a general purpose rotating drum mounted in line or perpendicular to the boom, or can be special function heads such as jack-hammer like spikes, compression fracture micro-wheel heads like those on larger tunnel boring machines, a slicer head like a gigantic chain saw for dicing up rock, or simple jaw-like buckets of traditional excavators.
- A forklift is a powered industrial truck used to lift and move materials over short distances. The forklift was developed in the early 20th century by various companies, including Clark, which made transmissions, and Yale and Towne Manufacturing, which made hoists.
- A forklift derives their power from two intertwining mechanisms: a pair of hydraulic cylinders and a pair of roller chain pulleys. When pressed, the handle activates the air pump, which draws in outside air through a filter and forces it into a tube leading to both hydraulic cylinders.

- A mobile concrete mixer is a batch plant mounted on a chassis–usually a truck or trailer–and carries unmixed material (sand, cement, coarse aggregates, water and any other materials or chemicals needed for more specialty applications) to a job site and mixed on a continuous or intermittent basis as required for fresh .
- Stone pavers are stones that are used to build flat, easily accessible surfaces for walkways, patios, driveways, and pool decks. Stone pavers (sometimes called "paving stones") usually come in squares or rectangles. Unlike concrete pavers and most brick pavers, they do not always come in identical sizes.
- A hoist is a device used for lifting or lowering a load by means of a drum or lift-wheel around which rope or chain wraps. It may be manually operated, electrically or pneumatically driven and may use chain, fiber or wire rope as its lifting medium.

Multiple Choice Questions

1. What is "LIDAR" ?
   (a) Light detection and rod for levelling (b) Light detection and ranging
   (c) Laser detection and ranging (d) None of these
2. "LIDAR" is a method of ......
   (a) foundation design (b) surveying
   (c) estimating (d) design of R.C.C. structures
3. "G.P.S." means ......
   (a) gravity pull system (b) gravity power system
   (c) global positioning system (d) global power presentation society
4. "G.P.S." is used for ......
   (a) map forests (b) helping farmers
   (c) navigation of aeroplane (d) all of the above
5. "G.P.S." is used for ......
   (a) location of a station on earth (b) military applications
   (c) surveying of land (d) all of the above
6. "G.I.S." stands for ......
   (a) global improvement system (b) global increment society
   (c) global information system (d) global intimation system
7. "G.I.S." is used for ......
   (a) geographical data storage (b) geographical data analysis
   (c) geographical data management (d) all of the above
8. "G.I.S." is not used for ......
   (a) studying geological features
   (b) seismic information
   (c) non-destructive testing
   (d) creating 3-d-display of geographical features
9. The science of making measurements of ground of land from photographs is called ......
   (a) photograph (b) tachometry
   (c) photogrammetry (d) photographical science and technology
10. "Drones" are controlled by ......
    (a) pilot inside it (b) petrol engine and steering
    (c) remote ground control system (d) pilot and helper sitting inside drone
11. A rod graduated in meters is used to determine difference in elevation is called ......
    (a) levelling staff (b) grad rod
    (c) ranging rod (d) cross-staff

12. "GCS" is called ......
    (a) ground controlled system (b) ground clearance system
    (c) ground smart system (d) none of these
13. Which is not a type of laser level ?
    (a) Dot laser level (b) Line laser level
    (c) Rotary laser level (d) Dumpy level
14. Equipments used for excavation, spreading of soil and hauling of rock, much etc. are called ......
    (a) earthmoving equipments (b) crushing equipments
    (c) concreting equipments (d) dumper and trucks
15. Which is not an earthmoving machine ?
    (a) J.C.B. (b) Hoe
    (c) clam shell (d) centrifugal pump
16. Bull-dozers are used to ......
    (a) level the ground (b) demolish the old structures
    (c) uprooting trees and shrubs (d) all of the above
17. Bridge girders are launched by ......
    (a) bulldozer (b) launching gantry
    (c) grader (d) three sheave pulley block
18. Which is not a part of crane ?
    (a) Hoist rope (b) Chain or wire ropes
    (c) Sheaves (d) Telescope
19. "TBM" stands for ......
    (a) tunnel balancing machine (b) tunnel boring machine
    (c) tunnel and brick making machine (d) total boring machine
20. "T.B.M." is also known as ......
    (a) mole (b) crane
    (c) hoist (d) grade rod
21. A forklift is a type of ......
    (a) truck (b) level
    (c) crane (d) mixer
22. Stone pavers are used for ......
    (a) walkways (b) drive ways
    (C) pool decks (d) all of the above
23. "Hoist" is a device used to ......
    (a) lift or lowering a load (b) weighing a load
    (c) excavation (d) surveying of land

Answers

1. (b)	2. (b)	3. (c)	4. (d)	5. (d)	6. (c)	7. (d)	8. (c)	9. (c)	10. (c)
11. (b)	12. (a)	13. (d)	14. (a)	15. (d)	16. (d)	17. (b)	18. (d)	19. (b)	20. (a)
21. (a)	22. (d)	23. (a)							

❑❑❑

Chapter **4**

Sustainable Resource Management

Weightage of Marks = 12, Teaching Hours = 08

Syllabus

4.1 Water resource management
- Principles and techniques

4.2 4R's in waste management
- Reduce, Reuse, Recycle and Recover
- Concept of Zero Waste

4.3 Reuse of waste in construction
- Fly Ash, Slag and Plastic

4.4 Renewable energy sources
- Solar energy, biomass, wind energy, ocean wave energy, Geo thermal energy and hybrid power system.

4.5 Energy Audit
- Necessity and methods.

4.6 Natural disaster management
- Flood, Earthquake, Tsunami, Volcanic Eruption, Hurricanes, Landslides.

4.7 Site Safety
- Necessity, principles, tools, techniques, laws, rules and regulations.

About this Chapter

At the end of this chapter, students will be able to:
- Elaborate the principle of water resource management.
- Adopt the suitable technique for water conservation.
- Justify the need of 4R principle in waste management.
- Recommend the use of waste for the given civil construction work.
- Suggest the appropriate safety technique for given site condition.

4.1 WATER RESOURCE MANAGEMENT

- Water is an essential resource for all life on the planet. Of the water resources on Earth only three percent of it is fresh and two-thirds of the freshwater is locked-up in ice caps and glaciers.
- Of the remaining one percent, a fifth is in remote, inaccessible areas and much seasonal rainfall in monsoonal deluges and floods cannot easily be used.
- As time advances, water is becoming scarcer and having access to clean, safe, drinking water is limited among countries.

- At present only about 0.08 percent of all the world's fresh water is exploited by mankind in ever increasing demand for sanitation, drinking, manufacturing, leisure and agriculture.
- Due to the small percentage of water remaining, optimizing the fresh water we have left from natural resources has been a continuous difficulty in several locations worldwide.
- Water resource management is the activity of planning, developing, distributing and managing the optimum use of water resources. It is a sub-set of water cycle management.
- Water is very essential for our survival. The field of water resources management will have to continue to adapt to the current and future issues facing the allocation of water. With the growing uncertainties of global climate change and the long term impacts of management actions, the decision-making will be even more difficult.
- It is likely that ongoing climate change will lead to situations that have not been encountered. As a result, alternative management strategies are sought for in order to avoid setbacks in the allocation of water resources.
- Ideally, water resource management planning has regard to all the competing demands for water and seeks to allocate water on an equitable basis to satisfy all uses and demands. As with other resource management, this is rarely possible in practice.
- One of the biggest concerns for our water-based resources in the future is the sustainability of the current and even future water resource allocation.
- As water becomes scarcer, the importance of how it is managed grows vastly. Finding a balance between what is needed by humans and what is needed in the environment is an important step in the sustainability of water resources.

4.1.1 Principles and techniques

Sources of Water

- There are mainly two sources of water :

  1. Surface Sources. 2. Sub-Surface Sources.

- **Surface Sources:** It is further divided as streams, rivers, lakes, ponds, natural or artificial reservoir.
- **Sub-Surface Sources:** These are divided as springs, wells, infiltration galleries, infiltration wells.

Factors on which the Selection of Source of Water Depends:

1. The source selected should provide water throughout the year to a city.
2. The quantity of water available from the source' selected should be adequate to meet the public demand.
3. The water source selected should be cheap.
4. The quality of water from source selected should be so good as n to require any expensive treatment.

- Site selection for an intake,

  1. The site should be near the treatment plant for minimising the cost of converging water.
  2. The site should provide maximum water from the source to the whole area.
  3. The site should not be located at the down stream. or in the vicinity of disposal of waste water.
  4. The site should be such that at the velocity of flow in the source is gentle.
  5. The site should be easily accessible and should be away from the effects of floods.
  6. The site should provide best quality in order to minimise the cost of purification.

Various sources of water are as below :

1. **Surface Sources:** (a) Streams, (b) Rivers, (c) Lakes, (d) Ponds, (e) Natural or Artificial Reservoir.
2. **Sub-Surface Sources:** (a) Springs, (b) Wells, (c) Infiltration Galleries, (d) Infiltration Wells.

1. Surface Sources: Surface sources are also known as ground sources. Water can be easily stored, naturally or artificially, on the earth surface.

(a) Streams: Small streams join to form river. The discharge of streams is more in rainy season than other seasons.

(b) Rivers: Rivers are the only source of surface water; which have maximum quantity of water; and can be easily taken into water supply.

(c) Lakes: Water from springs and streams flows towards basins and thus the lakes are formed. The lakes which are situated at high altitudes contain mostly pure water and can be used without any treatment.

(d) Ponds: Ponds are depression in plains; in which water is collected during rainy season. Ponds are mostly found in small villages.

(e) Natural or Artificial Reservoir: These reservoirs are used to supply water for irrigation facilities and for the community needs. Reservoirs also regulate the floods. As the water is stored in reservoir for long time, suspended impurities in it settle down.

2. Sub-Surface Sources: There are also known as underground sources. A portion of rainfall infiltrates in the ground and stored as sub-surface water at hard stratum.

(a) Springs: Underground water reappears on the earth surface in the form of springs. Springs can supply very little quantity of water; so inadequate for water supply.

(b) Wells: Vertical hole dug or drilled under the ground to tap the subsoil water is called well. Wells may be shallow or deep. The wells must be dug minimum 20 mts. away from the source of contamination.

(c) Infiltration Galleries: These are also called porous pipe galleries. The ground water existing in large quantity over a wide area can be collected by laying porous pipes in the full area at some places. The pipes are laid in a slope such that they can bring the water, where a well is constructed.

(d) Infiltration Wells: The shallow wells constructed under beds of rivers are called infiltration wells. These are provided to increase the yield of a well during summer.

Springs:

- Springs are the outcomes of underground water which ofton appears as small water holes at river banks or foot of hills.

Types of Springs:

1. **Gravity Springs:** The underground water level rises and the water outflows through the side of natural valley or a depression is called a gravity spring. The water table fluctuates regularly and the quantity of water in gravity springs becomes uncertain. The flow of water may be reduced to zero.
2. **Artesian Springs:** These are the wells from which water, flows automatically under pressure. This water can be stored in water tanks. These are mostly found in hilly regions.

Intakes:

- Intakes are the structures consisting the opening through which the raw water from river, lake and canal enters and carried upto a sump well by means of pipes, and water from the sump well is then pumped to water treatment units through the rising mains Intakes are used to collect water from various sources.

Types of Intakes:

1. **Lake Intakes:** These intakes are constructed in the bed of the lake below the water-level to collect water in dry season also. It consists of a pipe laid in the bed of the river, one end of which is fitted with bell mouth opening covered with mesh and protected by a concrete crib. The water is collected at the bank and flows under gravity in a sump well and then pumped to the water treatment plants.

2. **River Intakes:** It is a circular masonry tower-well of 4 to 7 m diameters. It is constructed along the river bank at the place from where water can be drawn in required quantity. The various river intakes are wet intake, dry intake and moval intake.

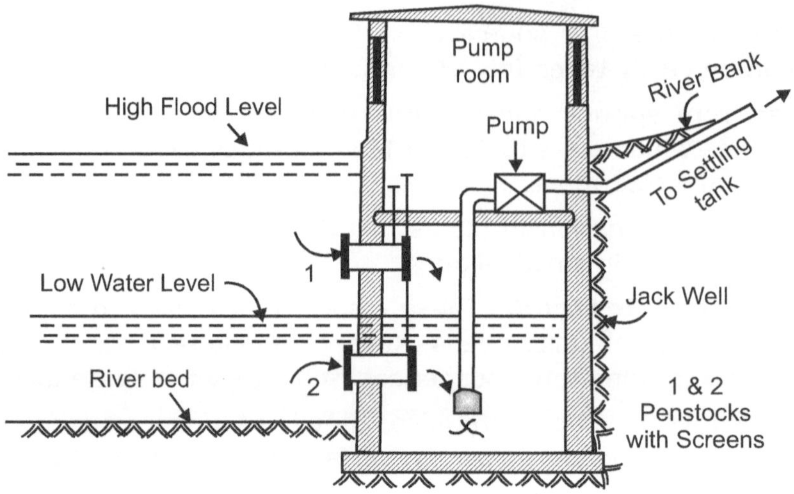

Fig. 4.1: River Intakes

3. **Reservoir Intakes:** These are used to collect water from dam reservoirs. It consists of an intake tower-well constructed on the slope of the dam at the place where required quantity of water can be collected even in the dry season. Intake pipes are fixed at different levels inside the intake-tower-well.

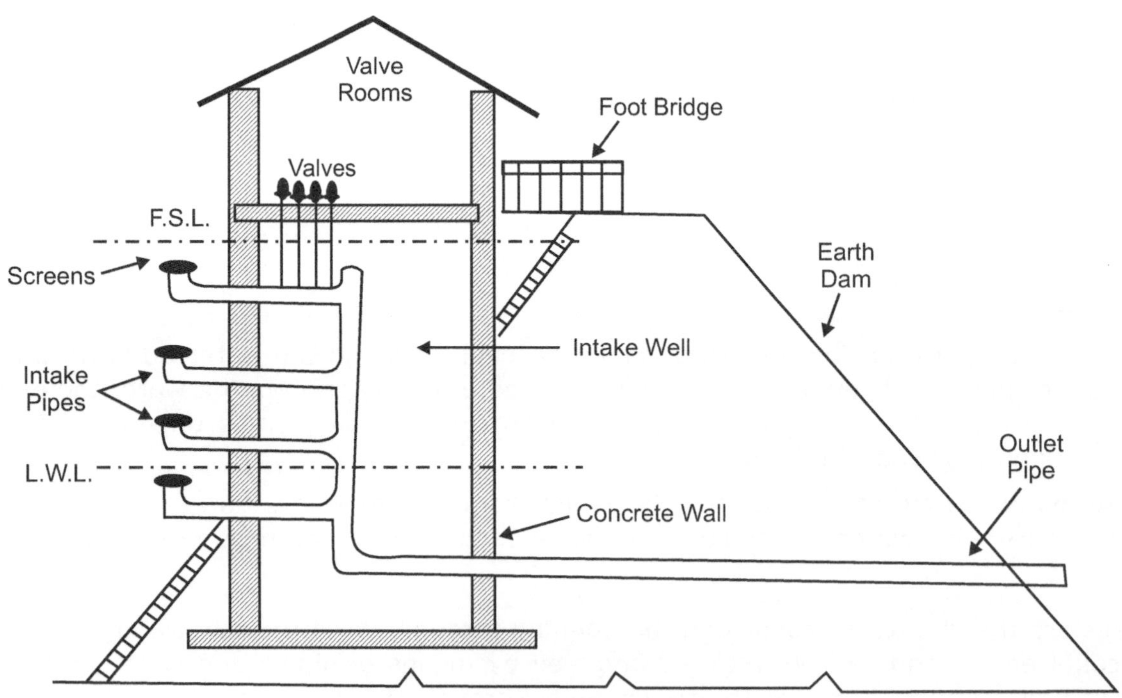

Fig. 4.2: Reservoir Intakes

4. **Canal Intakes:** Canal intake is the structure constructed on the canal bank. It consists of a pipe placed in brick masonary chamber constructed partly on the bank of the canal. The outlet pipe carries the water to the other wide of canal bank from where water is taken to the water treatment plant.

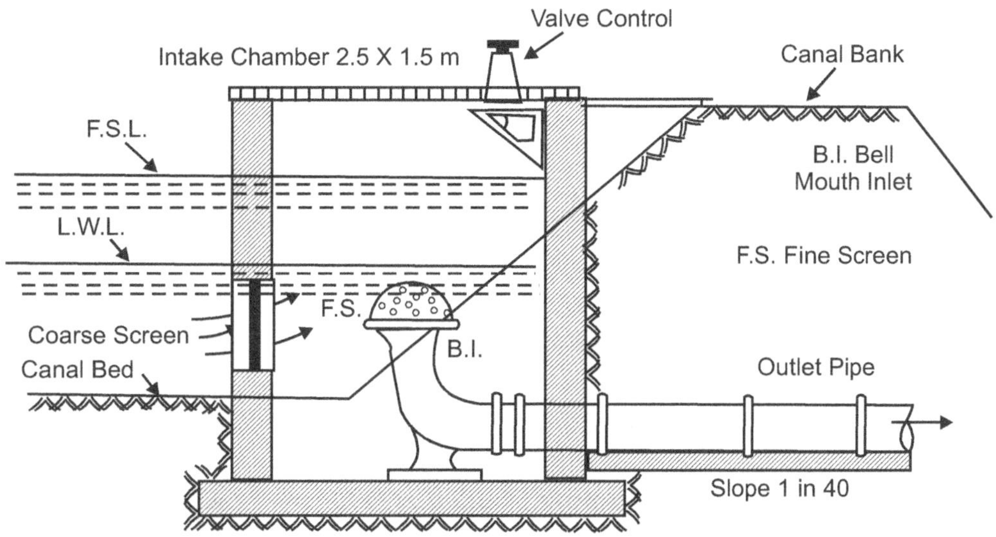

Fig. 4.3: Canal Intakes

4.1.2 Storage of Water

- Service reservoir is constructed to furnish storage to meet the fluctuating demands and to serve as balancing tanks to maintain the pressure during the hours of greatest demands.
- They meet emergency demands for extinguishing fire and failure of pumps etc.
- It allows the pump to work with normal rates and permits rest also.
- The storage is necessary if the hourly rate of peak demand exceeds the same pumping rate.
- The capacity of reservoir depends upon a number of factors as variation between maximum and minimum demands, pumping capacity and the number of hours the pumps arc designed to work and the stand-by arrangements.
- Generally their capacity is equal to 3 to 12 hours supply of water for the city, Service reservoirs should be located at the head of the distribution district, but the best site is the highest point in the district as regards the location.
- There are two types of service reservoirs, ground service reservoir and elevated service reservoir.

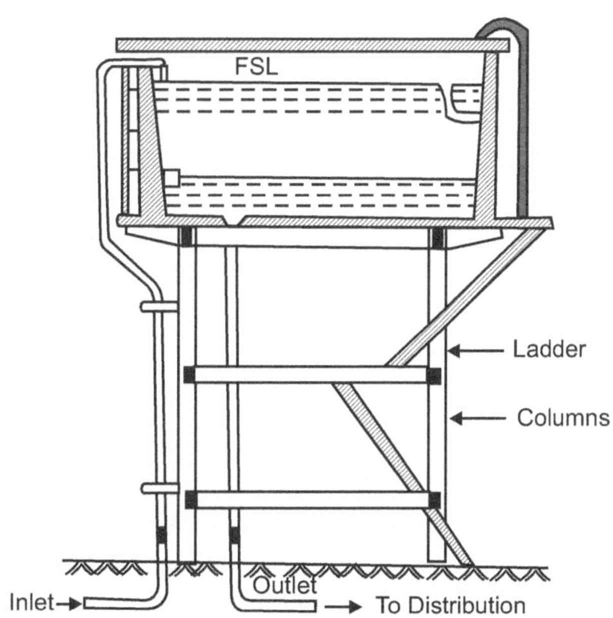

Fig. 4.4: Elevated Service Reservoir (E.S.R.)

Elevated Service Reservoir (E.S.R.)

Purpose:

- To provide enough capacity for balancing and equalizing reserve, breakdown reserve and fire reserve.
- To provide necessary head or elevation so that water will flow under gravity through the network of distribution system and water will reach the farthest consumer with sufficient pressure.

Working:

- The distributing areas which are not covered by gravity system of distribution these reservoirs are, constructed over there.
- These are constructed at an elevation from ground level. This reservoir consists of an inlet pipe through which treated water enters to the tank.
- Water is collected and stored in the tank. Outlet pipe is provided for exit of water which is connected to the distribution system.
- Water flows under head equal to the height of tank above ground level. Overflow pipe for the exit of water above full supply is provided.
- Wash out pipe is provided to remove the water during cleaning. Water level indicator, ventilators and manholes are also provided.
- Every pipe is provided with suitable valve. Ladder is provided to reach the top of reservoir.

Location:

- It is situated at the head of the distribution zone.
- It is located at the centre of area under its control in case of radial distribution.

Hot and Cold Water Supply System

- Plastic pipes are widely used in domestic hot and cold water supply and central heating systems. Their flexibility offers the benefits of easy handling and installation.
- They are available in both straight lengths and coils, which enables threading around obstructions and through joists, reducing the number of joints required and the potential for leaks.
- Where joints are required, the extensive range of fittings available for each pipe system enables reliable, watertight joints to be made quickly and easily.
- Cold water supply – pipework carrying water for domestic purposes including drinking, washing and operation of sanitary appliances.
- Hot water supply – pipework carrying heated water for domestic purposes including washing.

Hot Water Supply System Requirements for Domestic Residential and Commercial Building:

- Hot water supply is the distribution of stored heated water to consumers. The water can be heated using fuels such as coal, oil, gas, electricity, solar energy and nuclear energy.
- In brief, the hot water supply system involves a process to produce hot water from the cold water supply by heating the water in an airtight boiler, resistant to pressure, and, stored in the cylinder.
- Examples of areas that use hot water are as follows:

Medical:

(a) Use in places for washing clothes, bed sheets, blankets and so on in order to remove germs.

(b) Use for the sterilization of medical equipment.

(c) Use in study and research testing room, medical laboratory and chemistry laboratory.

Manufacturing:

(a) Use in palm oil mill where the hot water is used to steam oil palm fruit.

(b) Use in textile and dyeing industries for dipping, dyeing and washing clothes.

(c) Use in food canning industry for food preparation.

Residential:

(a) Use for cooking, drinking water and cleaning.

(b) Use for laundry.

(c) Use for heating the interior of the house in cold climate countries.

Tourism:

(a) Need in sauna, spa and guest rooms.

(b) Use for laundry, cleaning utensils in hotels kitchen and food preparation.

Water Heating Method:

There are two water-heating methods often used namely:

I. Electric Power:

(a) Systems that use electricity for heating water comprise a cylinder in which appropriate equipment will heat the water. The cylinders are the common galvanized steel or copper cylinder.

(b) The cold-water storage tanks or the service pipe supplies the cold water.

(c) The heater combines the water heater with hot water storage tank. The design is for use in large buildings.

II. Solar Energy:

(a) The use of solar energy is just for an additional energy or as aid to heat the water.

(b) The use of solar energy together with electric heater.

(c) The cylinder gathers the hot water from the collectors before sending to other heating tank.

(d) The heating tank will reheat the water before sending to the user.

(e) It can save the cost of electricity.

(f) A high initial cost for using cylinder and tank.

Direct Hot Water Supply System:

I. Using Cylinder:

(a) When the water runs through it. It heats the water immediately and does not store the water. Its size is small same like water heater and popular among users.

(b) It is equipped with a pressure switch, the function is to turn-off the heating element when no water is passing through.

(c) Has the 'thermal cut out switch' function as safety if the water exceeds the heat level.

(d) Usually, it connects directly to the main pipe of the cold water supply.

II. Using Tank:

(a) Electricity or gas generates the system. Electric heater tank consists of a cylinder with the appropriate equipment to heat the water. The type of cylinder is the common galvanized steel or copper cylinder.

(b) The cold-water storage tanks or the service pipe supplies the cold water

(c) The heater combines the water heater with hot water storage tank. The design is for use in large buildings.

Concepts of a rain water harvesting

- In every building position, number and sizes of doors and windows should be in such a way that sufficient day light free from glare comes from the right direction.
- Sufficient daylight should reach in each and every room of the building, to create pleasing environment to work and to live.
- The designer should also keep in mind the lighting requirements with heat insulation of the building.
- The purpose of design on natural light is to tackle the problem of solar heating and glare to the interior building from direct sunlight.
- Design with excellent natural lighting or day lighting is important to gain comfort level for healthy environment.
- Even though the precipitation cycle may fluctuate from year-to-year and place to place, the rain water's environmental advantage and purity over other water options makes it the first choice as a source of water.
- We all are dependent on rain water as an only source of water on this planet.
- Collecting the rain that falls on a building roofs and using the same for various purposes is a simple concept. Since the rain we harvest is independent of any centralized system, we are promoting self-sufficiency a helping to tap this essential and precious resource.
- The collection of rain water not only leads to conservation of water but also energy, since the energy not required to operate a centralized water system designed to treat and pump water over a vast service area.
- Policymakers would have to reconsider present assumptions regarding impervious cover and consequent run-off management strategies when rain water harvesting systems are installed.
- A compelling advantage of rain water over other water sources is that, it is one of the purest sources of water available.
- Rain water quality almost always exceeds that of ground or surface water as it does not come into primary water
- The quality of rain water is an overriding incentive for people to choose rain water as their source, or for specific uses such as watering houseplants and gardens.
- However, quality of rain water can be influenced by characteristics of area where it falls, since localized industrial emissions affect its purity.
- Thus, rain water falling in non-industrialized areas can be superior to that in cities dominated by heavy industry or in agricultural regions where crop dusting is prevalent.
- Rain water is soft water and can significantly reduce the quantity of detergents and soaps needed for cleaning, as compared to typical municipal water.
- In addition, soap scum and hardness deposits disappear and the need for a water softener, often an expensive requirement for well water systems, is eliminated.
- Water heaters and pipes will be free of deposits caused by hard water and will last longer.
- Rain water's purity also makes it an attractive water source for certain industries for which pure water is a requirement. Thus, industries such as computer microchip manufacturing and photographic processing would certainly benefit from this source of water.
- For our water requirement we entirely depend upon rivers, lakes and ground water. However, rain is the ultimate source that feeds all these sources.
- Rain water harvesting means to make optimum use of rain water at the place where it falls i.e. conserve it and not allow to drain away and cause floods elsewhere.
- **The rain water harvesting may be defined as, "the technique of collection and storage of rain water at surface or in sub-surface aquifer before it is lost as surface run-off".**

4.2 4R'S IN WASTE MANAGEMENT

4.2.1 Reduce

- REDUCE is to limit the amount of waste you create in the first place. This includes buying products with less packaging.

4.2.2 Reuse

- REUSE means to use something again that you would normally throw away (for example, Glass jar for food or plastic bags for bin liners.)
- One person's waste material can be useful material for others.
- Effort should therefore be made to encourage collection of such reusable material through waste collectors, waste producers, NGO's and private sector, instead of allowing reusable waste to land upon the disposal sites.
- Bottles, cans, tins, drums and cartons can be reused.
- Reuse corrugated moving boxes internally. Encourage employees to reuse office materials rather than purchase new ones.

4.2.3 Recycle

- Recycle means the product goes through a mechanical process to change its form. This is only recommended when reducing and reusing are not possible.
- In the age of excessive packaging materials being used, a lot of recyclable waste material is generated.
- All-out efforts are necessary to retrieve recyclable material from the households, shops and establishments and fed to the recycling industries through intermediaries such as waste purchasers, waste collectors/NGO's etc.

4.2.4 Recover

- Recovery of solid waste means recovering material from solid waste to useful purposes. Composting is a way to return nutrient back into the environment by allowing micro-organisms to turn the waste into manure.
- Recover is to convert waste into resources (such as electricity, heat, compost and fuel) through thermal and biological means.
- Resource Recovery occurs after reduce, reuse and recycle has been attempted.

4.2.5 Concept of Zero Waste

- **Zero Waste:** The conservation of all resources by means of responsible production, consumption, reuse, and recovery of all products, packaging, and materials, without burning them, and without discharges to land, water, or air that threaten the environment or human health.
- Zero Waste is a set of principles focused on waste prevention that encourages the redesign of resource life cycles so that all products are reused. The goal is for no trash to be sent to landfills, incinerators, or the ocean.
- Currently, only 9% of plastic is actually recycled.
- Zero Waste refers to waste prevention as opposed to end-of-pipe waste management.
- It is a whole systems approach that aims for a massive change in the way materials flow through society, resulting in no waste.
- Zero waste encompasses more than eliminating waste through recycling and reuse, it focuses on restructuring production and distribution systems to reduce waste.
- Zero waste is more of a goal or ideal rather than a hard target. Zero Waste provides guiding principles for continually working towards eliminating wastes.
- Advocates expect that government regulation is needed to influence industrial choices over product and packaging design, manufacturing processes, and material selection.

4.3 REUSE OF WASTE IN CONSTRUCTION

- The different industrial wastes like flyash, red mud, silica fumes and copper slag have replaced the traditional construction materials cement, sand, coarse aggregate. These industries wastes are valuable by-products that reduce the environment pollution.

4.3.1 Fly Ash

- The burning of harder older anthracite and bituminous coal produces Class F fly-ash. This fly-ash is pozzolonic in nature containing less than 20% lime. Hence, together with portland cement, quick lime or hydrated lime and water produces cementitious compounds. Addition of water glass to it, Class F fly-ash can form geopolymers.
- Class C Fly-ash : It is produced from burning younger ignite or sub-bituminous coal. In presence of water, it hardens and gains strength over time. It is used in making fly-ash bricks. Bricks and payers are used in commercial market.
- It is used in concrete production, as a substitute for portland cement and sand.
- It is used for embankments and road construction.
- It is used for production of grout and flowable fill for crack filler in buildings.
- It is used for waste stabilization and solidification.
- It is used in cement clinker production as a substitute for clay.
- It is used for land reclamation.
- It is used for stabilization of soft soils.
- It is used for road sub-base construction.
- It is used as a substitute for aggregate in brick production.
- It is used as a mineral filler in asphaltic concrete.
- It is used for the production of roller compacted concrete.

4.3.2 Slag

- Slag is used in making an aggregate having applications in granular base, embankments, engineered fill, in highways and hot-mix asphalt pavements.
- It is used as a concrete aggregate in construction of bridges.
- It is used in concrete masonry due to its desirable properties such as light weight and high fire resistance.
- It is used as an aggregate with asphalt for construction of damp-proof course and water proofing.
- It is used as a insulation in mineral wool, roof aggregate etc. in rail road ballast etc.

4.3.3 Plastic

- Plastic bags are popular with consumers and retailers as they are a functional, lightweight, strong, cheap, and hygienic way to transport food and other products.
- Most of these go to landfill and garbage heaps after they are used, and some are recycled.
- Once littered, plastic bags can find their way on to our streets, parks and into our waterways.
- Although plastic bags make-up only a small percentage of all litter, the impact of these bags is nevertheless significant. Plastic bags create visual pollution problems and can have harmful effects on aquatic and terrestrial animals. Plastic bags are particularly noticeable components of the litter stream due to their size and can take a long time to fully break down.
- Many carry bags end up as unsightly litter in trees, streets, parks and gardens which, besides being ugly, can kill birds, small mammals and other creatures.
- Bags that make it to the ocean may be eaten by sea turtles and marine mammals, who mistake them for jellyfish, with disastrous consequences.

- In developed countries billion bags are thrown away every year, most of which are used only once before disposal. The biggest problem with plastic bags is that they do not readily break down in the environment. It has been found that, the average plastic carrier bag is used for five minutes, but takes 500 years to decompose.
- Reuse is a step-up from recycling. It diverts plastic and takes pressure-off the recycling services. In fact, reuse is the middle-man between reduce and recycle, and some would be surprised at how many opportunities for reuse there really are.
- One can reuse plastic-produce bags for sandwiches, plastic grocery bags for small trash bags, and re-use plastic silverware. Most people skip this step and go directly to recycling, but reusing plastics can reduce the demand for new plastics to be created.
- For instance, since refillable plastic containers can be reused for many times, container reuse can lead to a substantial reduction in the demand for disposable plastic and reduced use of materials and energy, with the consequent reduced environmental impacts.
- Recycling and re-utilization of waste plastics have several advantages.
- It leads to a reduction of the use of virgin materials and of the use of energy, thus also a reduction of carbon dioxide emissions.

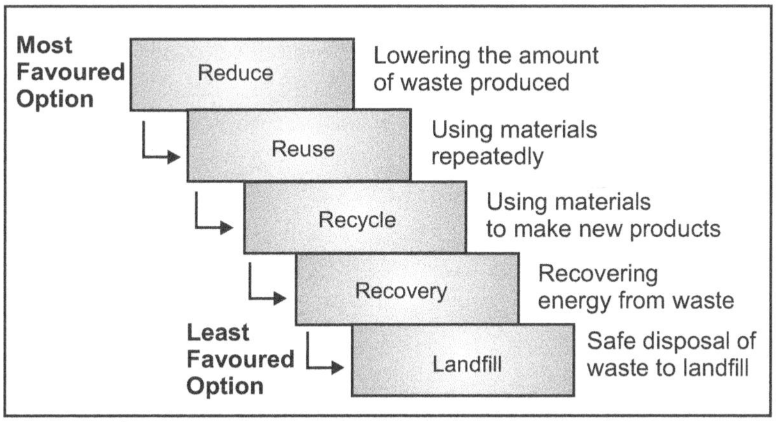

Fig. 4.5: Management of Plastic Waste

4.4 RENEWABLE ENERGY SOURCES

Renewable energy:

- Energy which comes from natural resources and is renewable, such as sunlight, wind, rain, tides and geothermal heat is known as renewable energy.
- It is derived from natural processes that are replenished constantly. In its various forms, it derives directly from the sun, or from heat generated deep within the earth. Included in the definition is electricity and heat generated from solar, wind, ocean, hydropower, biomass, geothermal resources, and biofuels and hydrogen derived from renewable resources.
- Today we primarily use fossil fuels to heat and power our homes and fuel our cars. It is convenient to use coal, oil, and natural gas for meeting our energy needs, but we have a limited supply of these fuels on the Earth. We are using them much more rapidly than they are being created. Eventually, they will run out.

4.4.1 Solar energy

- Solar energy is the energy derived from the sun through solar radiation. Solar powered electrical generation relies on photo-voltaics and heat engines.
- Solar applications includes space heating and cooling through solar architecture, daylighting, solar hot water, solar cooking, and high temperature process heat for industrial purposes.
- Solar technologies tap directly into the infinite power of the sun and use that energy to produce heat, light, and power.
- Passive Solar Lighting and Heating People have used the sun to heat and light their homes for centuries. Ancient Native Americans built their dwellings directly into south-facing cliff walls because they knew the sun travels low across the southern sky in the Northern Hemisphere during the winter.
- They also knew the massive rock of the cliff would absorb heat in winter and protect against wind and snow. At the same time, the cliff dwelling design blocked sunlight during the summer, when the sun is higher in the sky, keeping their dwellings cool.
- The modern version of this sun-welcoming design is called passive solar because no pumps, fans, or other mechanical devices are used. Its most basic features include large, south-facing windows that fill the home with natural sunlight, and dark tile or brick floors that store the sun's heat and release it back into the home at night.
- In the summer, when the sun is higher in the sky, window overhangs block direct sunlight, which keeps the house cool. Tile and brick floors also remain cool during the summer. Passive solar design combined with energy efficiency will go even further.
- Energy-efficient features such as energy saving windows and appliances, along with good insulation and weather stripping, can make a huge difference in energy and cost savings. Solar Water Heating Solar energy can be used to heat water for your home or your swimming pool.
- Most solar water-heating systems consist of a solar collector and a water storage tank. Solar water-heating systems use collectors, generally mounted on a south-facing roof, to heat either water or a heat-transfer fluid, such as non-toxic antifreeze. The heated water is then stored in a water tank similar to one used in a conventional gas or electric water-heating system.
- Solar Electricity Solar electricity or photovoltaic (PV) technology converts sunlight directly into electricity. Solar electricity has been a prime source of power for space vehicles since the inception of the space program.
- It has also been used to power small electronics and rural and agricultural applications for three decades.
- During the last decade, a strong solar electric market has emerged for powering urban grid-connected homes and buildings as a result of advances in solar technology along with global changes in electric industry restructuring.
- Although many types of solar electric systems are available today, they all consist of basically three main items: modules that convert sunlight into electricity; inverters that convert that electricity into alternating current so it can be used by most household appliances; and possibly or sometimes batteries that store excess electricity produced by the system.
- The remainder of the system comprises equipment such as wiring, circuit breakers, and support structures.

4.4.2 Biomass

- Biomass (plant material) is a renewable energy source. Through the process of photosynthesis, plants capture the sun's energy.
- When the plants are burnt, they release the sun's energy they contain. In this way, biomass functions as a sort of natural battery for storing solar energy.
- As long as biomass is produced sustainably, with only as much used as is grown, the battery will last indefinitely.

- Biomass is second to hydropower as a leader in renewable energy production. Biomass has an existing capacity of over 7,000 MW.
- Biomass as a fuel consists of organic matter such as industrial waste, agricultural waste, wood, and bark.
- Biomass can be burned directly in specially designed power plants, or used to replace up to 15% of coal as a fuel in ordinary power plants.
- Biomass burns cleaner than coal because it has less sulfur, which means less sulfur dioxide will be emitted into the atmosphere. Biomass can also be used indirectly, since it produces methane gas as it decays or through a modern process called gasification.
- Methane can produce power by burning in a boiler to create steam to drive steam turbines or through internal combustion in gas turbines and reciprocating engines.
- The largest use of biomass energy in Virginia is the forest products industry. Furniture plants, sawmills, and paper mills usually burn their wood waste to produce heat and electricity. Many homeowners use firewood or pellets for winter heat.

4.4.3 Wind energy

- For hundreds of years, people have used windmills to harness the wind's energy.
- Today's wind turbines, which operate differently from windmills, are a much more efficient technology.
- Wind turbine technology may look simple: the wind spins turbine blades around a central hub; the hub is connected to a shaft, which powers a generator to make electricity.
- However, turbines are highly sophisticated power systems that capture the wind's energy by means of new blade designs or airfoils.
- Modern, mechanical drive systems, combined with advanced generators, convert that energy into electricity.
- Airflows can be used to run wind turbines. Modern wind turbines range from around 600 kW to 5 MW of rated power, although turbines with rated output of 1.5-3 MW have become the most common for commercial, use; the power output of a turbine is a function of the cube of the wind speed, so as wind speed increases, power output increases dramatically.
- Areas where winds are stronger and more constant, such as offshore and high altitude sites, are preferred locations for wind farms.
- Wind energy is the kinetic energy associated with the movement of atmospheric air which can be harnessed to generate power.
- It is conversion of wind energy into a useful form of energy, such as using: wind turbines to make electricity, windmills for mechanical power, wind pumps for water pumping or drainage, or sails to propel ships.
- A large wind farm may consist of several hundred individual wind turbines which are connected to the electric power transmission network.
- Offshore wind farms can harness more frequent and powerful winds than are available to land-based installations and have less visual impact on the landscape but construction costs are considerably higher.
- Small onshore wind facilities are used to provide electricity to isolated locations and utility companies increasingly buy back surplus electricity 91 produced by small domestic wind turbines.

Merits of wind energy

1. It is a non-polluting and environment friendly source of energy.
2. It is a renewable energy available at free of cost.
3. Power generation is cheaper with nil recurring expenses.
4. Wind mills are suitable to erect at on shore, remote and rural areas where wind blows with required intensity.
5. Favorable in geographic locations which are away from cities.

Demerits of wind energy

1. Manufacturing, installation and maintenance are costly.
2. Wind turbine design, manufacturing, installation is complex due to varying atmospheric conditions.
3. Wind power doesn't suitable for large scale generation.
4. A major disadvantage of wind turbines is the noise pollution that it creates.

4.4.4 Ocean wave energy

- The ocean can produce two types of energy: thermal energy from the sun's heat, and mechanical energy from the tides and waves.
- Ocean thermal energy can be used for many applications, including electricity generation.
- Electricity conversion systems use either the warm surface water or boil the seawater to turn a turbine, which activates a generator.
- The electricity conversion of both tidal and wave energy usually involves mechanical devices.
- A dam is typically used to convert tidal energy into electricity by forcing the water through turbines, activating a generator.
- Meanwhile, wave energy uses mechanical power to directly activate a generator, or to transfer to a working fluid, water, or air, which then drives a turbine/generator

4.4.5 Geo thermal energy

- Geothermal energy is thermal energy generated and stored in the Earth.
- Thermal energy is the energy that determines the temperature of matter.
- Earth's geothermal energy originates from the original formation of the planet (20%) and from radioactive decay of minerals (80%).
- The geothermal gradient, which is the difference in temperature between the core of the planet and its surface, drives a continuous conduction of thermal energy in the form of heat from the core to the surface.
- The adjective geothermal originates from the Greek roots geo, meaning earth, and thermos, meaning heat.
- The Earth's core, 4,000 miles below the surface, can reach temperatures of 9000° F. This heat - geothermal energy – flows outward from the core, heating the surrounding area, which can form underground reservoirs of hot water and steam.
- These reservoirs can be tapped for a variety of uses, such as to generate electricity or heat buildings. By using geothermal heat pumps (GHPs), we can even take advantage of the shallow ground's stable temperature for heating and cooling buildings.
- **Geothermal Electricity Production**: Geothermal power plants access the underground steam or hot water from wells drilled a mile or more into the earth. The steam or hot water is piped-up from the well to drive a conventional steam turbine, which powers an electric generator. Typically, the water is then returned to the ground to recharge the reservoir and complete the renewable energy cycle.

Advantages of Geothermal Energy

- Mostly emissions and pollution-free
- Though initial drilling for power plants may be expensive, operation costs ar relatively inexpensive
- Has the potential to be the cheapest and most consistent/predictable type o renewable energy
- Massive and well-distributed resource
- It reduces the global warming and the growth of greenhouse gas emission.
- Baseload resource: can produce power 24 hours a day, unlike other renewable energy resources such as solar and wind energy

Disadvantages of Geothermal Energy

- Drilling for geothermal energy can be expensive.
- Rocks could lose heat after a few decades of drilling and new wells would have to be installed at new sites.

- The extracted fluid might contain toxic elements such as hydrogen sulfide (H_2S) and carbon dioxide (CO_2), mercury and other elements, this environment impacts can be provided by using a close system. Could potentially trigger earthquakes; in Basel, Switzerland, an earthquake occurred near the drilling site Tgcchna ANY krirSlatively new: has trouble attracting large-scale funding.

4.4.6 Hybrid power system

- A hybrid energy system, or hybrid power, usually consists of two or more renewable energy sources used together to provide increased system efficiency as well as greater balance in energy supply.
- For example, consider a load of 100% power supply and there is no renewable system to fulfill this need, so two or more renewable energy system can be combined.

  For example, 60% from a biomass system, 20% from wind system and the remainder from fuel cells. Thus combining all these renewable energy systems may provide 100% of the power and energy requirements for the load, such as a home or business.
- Another example of a hybrid energy system is a photovoltaic array coupled with a wind turbine. This would create more output from the wind turbine during the winter, whereas during the summer, the solar panels would produce their peak output. Hybrid energy systems often yield greater economic and environmental returns than wind, solar, geothermal or tri-generation standalone systems by themselves.

4.5 ENERGY AUDIT

- Energy Audit is the key to a systematic approach for decision-making in the area of energy management.
- It attempts to balance the total energy inputs with its use, and serves to identify all the energy streams in a facility.
- It quantifies energy usage according to its discrete functions.
- Industrial energy audit is an effective tool in defining and pursuing comprehensive energy management programme.
- As per the Energy Conservation Act, 2001, Energy Audit is defined as "the verification, monitoring and analysis of use of energy including submission of technical report containing recommendations for improving energy efficiency with cost benefit analysis and an action plan to reduce energy consumption".
- An industrial energy audit is a process that facilities energy usage patterns, equipment efficiency, and overall building efficiency are determined in order to propose energy efficiency measures.
- The result of a successful energy audit is decreased energy consumption, reduced raw material usage and increased quality of the end product.
- The data collected by an energy auditor is the basis on which the energy efficiency suggestions will be created.
- The implementation of these measures will reduce manufacturing costs and also the negative effects on the environment.

4.5.1 Necessity

- In any industry, the three top operating expenses are often found to be energy (both electrical and thermal), labour and materials.
- If one were to relate to the manageability of the cost or potential cost savings in each of the above components, energy would invariably emerge as a top ranker, and thus energy management function constitutes a strategic area for cost reduction.

- Energy Audit will help to understand more about the ways energy and fuel are used in any industry, and help in identifying the areas where waste can occur and where scope for improvement exists.
- The Energy Audit would give a positive orientation to the energy cost reduction, preventive maintenance and quality control programmes which are vital for production and utility activities.
- Such an audit programme will help to keep focus on variations which occur in the energy costs, availability and reliability of supply of energy, decide on appropriate energy mix, identify energy conservation technologies, retrofit for energy conservation equipment etc.
- In general, Energy Audit is the translation of conservation ideas into realities, by lending technically feasible solutions with economic and other organizational considerations within a specified time frame.
- The primary objective of Energy Audit is to determine ways to reduce energy consumption per unit of product output or to lower operating costs.
- Energy Audit provides a "bench-mark" (Reference point) for managing energy in the organization and also provides the basis for planning a more effective use of energy throughout the organization.

4.5.2 Methods

- The type of Energy Audit to be performed depends on :
  - Function and type of industry,
  - Depth to which final audit is needed, and
  - Potential and magnitude of cost reduction desired.
- Thus Energy Audit can be classified into the following two types.
  (i) Preliminary Audit.
  (ii) Detailed Audit.

4.5.2.1 Preliminary Energy Audit Methodology

- Preliminary energy audit is a relatively quick exercise to:
  - Establish energy consumption in the organization.
  - Estimate the scope for saving.
  - Identify the most likely (and the easiest areas for attention.
  - Identify immediate (especially no-/low-cost) improvements/savings.
  - Set a 'reference point'.
  - Identify areas for more detailed study/measurement.
  - Preliminary energy audit uses existing, or easily obtained data.

Detailed Energy Audit Methodology

- A comprehensive audit provides a detailed energy project implementation plan for a facility, since it evaluates all major energy using systems.
- This type of audit offers the most accurate estimate of energy savings and cost. It considers the interactive effects of all projects, accounts for the energy use of all major equipment, and includes detailed energy cost saving calculations and project cost.
- In a comprehensive audit, one of the key elements is the energy balance. This is based on an inventory of energy using systems, assumptions of current operating conditions and calculations of energy use.
- This estimated use is then compared to utility bill charges.
- Detailed energy auditing is carried out in three phases: Phase I, II and III.

  Phase I - Pre Audit Phase.

  Phase II - Audit Phase.

  Phase III - Post Audit Phase.

- **Phase I - Pre Audit Phase Activities:** A structured methodology to carry out an energy audit is necessary for efficient working. An initial study of the site should always be carried out, as the planning of the procedures necessary for an audit is most important.
- **Phase II - Detailed Energy Audit Activities:** Depending on the nature and complexity of the site, a comprehensive audit can take from several weeks to several months to complete. Detailed studies to establish, and investigate, energy and material balances for specific plant departments or items of process equipment are carried out. Whenever possible, checks of plant operations are carried out over extended periods of time, at nights and at weekends as well as during normal daytime working hours, to ensure that nothing is overlooked.

4.6 NATURAL DISASTER MANAGEMENT

- An earthquake is disastrous only when it affects people, their properties and activities. Thus, disaster occurs only when hazards and vulnerability meet.
- But it is also to be noted that with greater capacity of the individual/community and environment to face these disasters, the impact of a hazard reduces. Therefore, we need to understand the three major components namely hazard, vulnerability and capacity with suitable examples to have a basic understanding of disaster management.
- Natural hazards are hazards which are caused because of natural phenomena (hazards with meteorological, geological or even biological origin).
- Examples of natural hazards are hurricanes, tsunamis, earthquake and volcanic eruption which are exclusively of natural origin. Landslides, floods, drought, fires are socio-natural hazards since their causes are both natural and manmade.

  For example, flooding may be caused because of heavy rains, landslide or blocking of drains with human waste.

4.6.1 Flood

- Flood is a state of high water level along a river channel or on the coast that leads to inundation of land, which is not usually submerged. Floods may happen gradually and also may take hours or even happen suddenly without any warning due to breach in the embankment, spill over, heavy rains etc.
- Every stream flows in a particular stage at a particular place during different intervals of time in a year such as a dry stage, half-full stage and bank full stage.
- This depends upon climate and precipitation in the catchment area.
- A flood occurs when the volume of water in the river becomes greater than bank-full stage: the extra water spills over the banks and spread in sheets all along and away from the banks governed by available slope. This condition is called flood.

Types of floods:

- According their duration flood can be divided into different categories:
  - **i) Slow-Onset Floods:** Slow onset floods usually last for a relatively longer period, it may last for one or more weeks, or even months.
  - **ii) Rapid-Onset Floods**: Rapid onset floods last for a relatively shorter period, they usually last for one or two days only.
  - **iii) Flash Floods:** Flash floods may occur within minutes or few hrs. after heavy rainfall, tropical storm, failure of dams or releases water from dams and it causes the greatest damages to the society.

Causes of floods:

- Heavy Rainfall: Due to the sudden heavy rainfall the water level of water resources will increases due to this flooding will occurred.
- Due to the heavy siltation of the river bed, reduces the water carrying capacity of the rivers.
- Blockage in the drains lead to flooding of the area.
- When landslides occurs due to that the blocking will occurred in the flow of stream.
- Due to tsunami, the water level will increase, indirectly flooding will occur.
- Due to the earthquake, volcanic eruption, there will be major failure in dam, structure, due to that flooding will be occur.

Effect:

- Loss of life, damage to buildings and other structures (bridge, road, canals)
- Damage to power transmission and sometimes power generation which will indirectly affect the loss of power.
- This includes losses of drinking water treatment and water supply, which may results in loss of drinking water or serve water contamination.
- Clear water combined with human sewage in flood raises the risk of water borne dieses (typhoid, cholera).
- Loss of communication system.

Warning:

- Flood forecasting and warning has been highly developed in the past two decades. With the advancement of technology such as satellite and remote-sensing equipment's flood waves can be tracked as the water and Water Resources Department.
- CWC (Central Water Commission) maintains close liaison with the administrative and state engineering agencies, local civil authorities to communicate advance warning for appropriate mitigation and preparedness measures.

Possible Risk Reduction Measures:

- Mapping of the flood prone areas is a primary step involved in reducing the risk of the region. Historical records give the indication of the flood inundation areas and the period of occurrence and the extent of the coverage. Warning can be issued looking into the earlier marked heights of the water levels in case of potential threat. In the coastal areas the tide levels and the land characteristics will determine the submergence areas. Flood hazard mapping will give the proper indication of water flow during floods.
- **Land use control** will reduce danger of life and property when waters inundate the floodplains and the coastal areas. The number of casualties is related to the population in the area at risk. In areas where people already have built their settlements, measures should be taken to relocate to better sites so as to reduce vulnerability. No major development should be permitted in the areas which are subjected to high flooding. Important facilities like hospitals, schools should be built in safe areas. In urban areas, water holding areas can be created like ponds, lakes or low-lying areas.
- Construction of engineered structures in the flood plains and strengthening of structures to withstand flood forces and seepage. The buildings should be constructed on an elevated area. If necessary build on stilts or platform.
- Flood Control aims to reduce flood damage. This can be done by decreasing the amount of runoff with the help of reforestation (to increase absorption could be a mitigation strategy in certain areas), protection of vegetation, clearing of debris from streams and other water holding areas, conservation of ponds and lakes etc. Flood Diversion include levees, embankments, dams and channel improvement. Dams can store water and can release water at a manageable rate. But failure of dams in earthquakes and operation of releasing the water can cause floods in the lower areas. Flood Proofing reduces the risk of damage. Measures include use of sand bags to keep flood water away, blocking or sealing of doors and windows of houses etc. Houses may be elevated by building on raised land. Buildings should be constructed away from water bodies.

Flood Management

- In India, systematic planning for flood management commenced with the Five Year Plans, particularly with the launching of National Programme of Flood Management in 1954.
- During the last 48 years, different methods of flood protection structural as well as non-structural have been adopted in different states depending upon the nature of the problem and local conditions.

- Structural measures include storage reservoirs, flood embankments, drainage channels, anti-erosion works, channel improvement works, detention basins etc. and non-structural measures include flood forecasting, flood plain zoning, flood proofing, disaster preparedness etc.
- The flood management measures undertaken so far have provided reasonable degree of protection to an area of 15.81 million hectares throughout the country.

4.6.2 Earthquake

- Earthquake is one of the most destructive natural hazard. They may occur at any time of the year, day or night, with sudden impact and little warning. They can destroy buildings and infrastructure in seconds, killing or injuring the inhabitants. Earthquakes not only destroy the entire habitation but may de-stabilize the government, economy and social structure of the country.
- **Definition:** It is the sudden shaking of the earth crust. The impact of an earthquake is sudden and there is hardly any warning, making it impossible to predict.
- Globally, earthquakes result in a loss of about 50,000 lives every year. Earthquakes over 5.5 magnitude on the Richter scale are progressively damaging to property and human life. However, there are many other factors that influences the damage pattern. Massive earthquakes generally occur near the junction of two tectonic plates, for example, along the Himalayan range, where the Indian plate goes below Eurasian plate. The Indian sub- continent situated on the boundaries of two continental plates is very prone to earthquakes. Some of the most intense earthquakes of the world have occurred in India. Fortunately, none of these have occurred in any of the major cities. According to latest seismic zoning map brought out by the Bureau of Indian Standard (BIS), over 65 percent of the country is prone to earthquake of intensity Modified Mercalli Intensity Scale (MSK) VII or more.
- India has been divided into four seismic zones according to the maximum intensity of earthquake expected. Of these, zone V is the most active which comprises of whole of Northeast India, the northern portion of Bihar, Uttarakhand, Himachal Pradesh, J and K, Gujarat and Andaman and Nicobar Islands. India has highly populous cities and the constructions in these cities are not earthquake resistant. Regulatory mechanisms are weak, thus any earthquake striking in one of these cities would turn into a major disaster. Six major earthquakes have struck different parts of India over a span of the last 15 years.

Cause of Earthquake :

- The earth's crust is a rocky layer of varying thickness ranging from a depth of about 10 kilometers under the sea to 65 kilometers under the continents. The crust is not one piece but consists of portions called 'plates' which vary in size from a few hundred to thousands of kilometers. The 'theory of plate tectonics' holds that the plates ride up on the more mobile mantle, and are driven by some yet unconfirmed mechanisms, perhaps thermal convection currents. When these plates contact each other, stress arises in the crust.
- These stresses can be classified according to the type of movement along the plate's boundaries:

  (a) pulling away from each other,

  (b) pushing against one another, and

  (c) sliding sideways relative to each other.
- All these movements are associated with earthquakes.

Characteristics

- Earthquake vibrations occur in a variety of frequencies and velocities. The actual rupture process may last for a few seconds to as long as one minute for a major earthquake. The ground shaking is caused by 'body waves' and 'surface wave'.
- **Body waves** (P and S waves) penetrate the body of the earth, vibrating fast. 'P' waves travel about 6 kilometers per hour and 'S' waves travel with a speed of 4 kilometers per hour.

- Surface waves vibrate the ground horizontally and vertically. These long period waves cause swaying of tall buildings and slight waves motion in bodies of water even at great distances from the epicenter.
- Earthquakes can be of three types based on the focal depth:
  Deep: 300 to 700 kms from the earth surface.
  Medium: 60 to 300 kms.
  Shallow: less than 60 kms.
- The deep focus earthquakes are rarely destructive because by the time the waves reach the surface the impact reduces.
- Shallow focus earthquakes are more common and are extremely damaging because of their proximity to the surface.

Measuring Earthquakes:

- Earthquakes can be described by the use of two distinctively different scales of measurement demonstrating magnitude and intensity. Earthquake magnitude or amount of energy released is determined by the use of a seismograph' which is an instrument that continuously records ground vibration. The scale was developed by a seismologist named Charles Richter. An earthquake with a magnitude 7.5 on the Richter scale releases 30 times the energy than one with 6.5 magnitudes. An earthquake of magnitude 3 is the smallest normally felt by humans. The largest earthquake that has been recorded with this system is 9.25 (Alaska, 1969 and Chile, 1960)

4.6.3 Tsunami

- The term Tsunami has been derived from a Japanese term Tsu meaning 'harbor' and nami meaning 'waves'.
- Tsunamis are popularly called tidal waves but they actually have nothing to do with the tides.
- These waves which often affect distant shores, originate by rapid displacement of water from the lake or the sea either by seismic activity, landslides, volcanic eruptions or large meteoroid impacts.
- Whatever the cause may be sea water is displaced with a violent motion and swells up, ultimately surging over land with great destructive power.
- The effects of a tsunami can be unnoticeable or even destructive.

Causes of a Tsunami

- The geological movements that cause tsunamis are produced in three major ways. The most common of these are fault movements on the sea floor, accompanied by an earth-quake.
- They release huge amount of energy and have the capacity to cross oceans. The degree of movement depends on how fast the earthquake occurs and how much water is displaced.
- Tsunamis and earthquakes happen after centuries of energy build-up within the earth. A tsunami (in Japanese 'tsu' means harbor and 'nami' means wave) is a series of water waves caused by the displacement of a large volume of a body of water, usually an ocean. In the Tamil language it is known as "Aazhi Peralai".
- Seismicity generated tsunamis are result of abrupt deformation of sea floor resulting vertical displacement of the overlying water.
- Earthquakes occurring beneath the sea level, the water above the reformed area are displaced from its equilibrium position.
- The release of energy produces tsunami waves which have small amplitude but a very long wavelength (often hundreds of kilometer long). It may be caused by non-seismic event also such as a landslide or impact of a meteor.
- The second most common cause of the tsunami is a landslide either occurring under water or originating above the sea and then plunging into the water.
- The largest tsunami ever produced by a landslide was in Lituya Bay, Alaska 1958. The massive rock slide produced a wave that reached a high water mark of 50-150 meters above the shoreline.

- The third major cause of tsunami is volcanic activity. The flank of a volcano located near the shore or under water may be uplifted or depressed similar to the action of a fault, or, the volcano may actually explode.
- In 1883, the violent explosion of the famous volcano, Krakotoa in Indonesia, produced tsunami measuring 40 meters which crushed upon Java and Sumatra. Over 36,000 people lost their lives in this tyrant waves.

Typical adverse effects:

- Local tsunami events or those less than 30 minutes from the source cause the majority of damage. The force of the water can raze everything in its path. It is normally the flooding effect of the tsunami that causes major destruction to the human settlements, roads and infrastructure thereby disrupting the normal functioning of the society.
- Withdrawal of the tsunami causes major damage. As the waves withdraw towards the ocean they sweep out the foundations of the buildings, the beaches get destroyed and the houses carried out to sea. Damage to ports and airports may prevent importation of needed food and medical supplies. Apart from the physical damage, there is a huge impact on the public health system. Deaths mainly occur because of drowning as water inundates homes. Many people get washed away or crushed by the giant waves and some are crushed by the debris, causes.
- There are very few evidences which show that tsunami flooding has caused large scale health problem.
- Availability of drinking water has always been a major problem in areas affected by a disaster. Sewage pipes may be damaged causing major sewage disposal problems. Open wells and other ground water may be contaminated by salt water and debris and sewage. Flooding in the locality may lead to crop loss, loss of livelihood like boats and nets, environmental degradation etc.

Possible risk reduction measures:

- While it is of course not possible to prevent a tsunami, in certain tsunami prone countries some measures have been taken to reduce the damage caused on shore. Japan has implemented an extensive programme of building tsunami walls of up to 4.5 m (13.5 ft) high in front of populated coastal areas. Other localities have built flood gates and channels to redirect the water from incoming tsunamis.
- Framed structures in the area. The wall may have succeeded in slowing down and moderating the height of the tsunami but it did not prevent major destruction and loss of life.
- Some other systematic measures to protect coastlines against tsunamis include Event of one kind of Disaster sometimes lead to a cascading effect and untold other disasters. The preparedness to respond to such unforeseen chain of events is difficult to visualize and plan and is going to be one of the challenges to managers engaged in DRR. Recent experiences of earthquake in Japan leading to Tsunami, fire and then nuclear emergency are an eye opener to those underestimating the might and fury of nature.
- On 11th March 2011, at 14:46 the northeastern part of Japan was triggered by a major earthquake of magnitude 9 with an epicenter at 130 km off the pacific coast of Tohoku region and depth of 24 km. As per the Japan Meteorological Agency (JMA), the seismic intensity of 7[in the Japanese scale of 1 to 7] was recorded in Kurihara city of Miyagi prefecture. Moreover, the intensity of 6+ was recorded in 28 cities and towns in Miyagi, Fukushima, Ibaraki, and Tochigi prefacture. It also included Wakuya town, Tome city and Osaki city. Intensity of 6 or weaker was observed in different parts of the country.

4.6.4 Volcanic Eruption

- Magma is present inside the earth. It has a lesser density than the rocks above it. So it comes up whenever there is a path of fracture available to it.
- It is the under pressure of the surrounding rocks so it comes out of a volcano as a jet.
- This eruption consists of lava, rocks, dust, ashes, soil and gases.
- It is very hot and spreads all along the slope of the volcano.

- It cases a great damage to the nearby habitations, property and the environment, toxic and hazardous gases pollute the environment.
- It all cases same as earthquake.
- Volcanoes may remain dormant for years and may start ejecting the eruptions suddenly. There should be some scientific study of each volcano so that some safe period is available.

4.6.5 Hurricanes

- Hurricane is a region of low atmospheric pressure surrounded by high atmospheric pressure resulting in swirling atmospheric disturbance accompanied by powerful winds blowing in anticlockwise direction in the Northern Hemisphere and in the clockwise direction in the Southern Hemisphere. They occur mainly in the tropical and temperate regions of the world.
- The yearly distribution of tropical hurricanes in the north Indian Ocean indicates large year-to-year variations in the frequency of cyclonic disturbances and tropical hurricanes, but no distinct periodicity.
- A severe super cyclonic storm with winds of upto 250 km/hour crossed the coast in Orissa on October 29, 1999. This may have been the worst hurricane of the country in the Orissa region and was responsible for as many as 10,000 deaths, rendering millions homeless and extensive damage to property and environment
- The major natural disaster that affects the coastal regions of India is hurricane and as India has a coastline of about 7516 kms, it is exposed to nearly 10 percent of the world's tropical hurricanes. About 71 percent of this area is in ten states (Gujarat, Maharashtra, Goa, Karnataka, Kerala, Tamil Nadu, Puducherry, Andhra Pradesh, Orissa and West Bengal). The islands of Andaman, Nicobar and Lakshadweep are also prone to hurricanes. On an average, about five or six tropical hurricanes form in the Bay of Bengal and Arabian sea and hit the coast every year.
- Out of these, two or three are severe. When a hurricane approaches to coast, a risk of serious loss or damage arises from severe winds, heavy rainfall, storm surges and river floods. The effect of a storm surge is most pronounced in wide and shallow bays exposed to hurricanes such as in the northern part of Bay of Bengal. On an average, five or six tropical hurricanes occur every year, of which two or three could be severe. Most hurricanes occur in the Bay of Bengal followed by those in the Arabian Sea and the ratio is approximately 4 : 1. The incidence of cyclonic storms, with wind speeds between 65 Km/h and 117 Km/h and severe cyclonic storm with wind speeds between 119 Km/h and 164 Km/h, reaching Tamil Nadu and Andhra Pradesh is high during the north east monsoon season i.e. October – December, whereas the highest annual number of storms, severe storms occur in the Orissa - West Bengal coast.

Indian Hurricanes :

- Hurricanes vary in frequency in various parts of the world. The 7516.6 kilometers long Indian coastline is the earth's most hurricane battered stretch of the world. Around 8 per cent of the total land area in India is prone to hurricanes.
- About two-third of the hurricanes that occur in the Indian coastline occur in the Bay of Bengal. The states which are generally affected in the east coast are West-Bengal, Orissa, Andhra Pradesh; Tamil Nadu and on the west coast Gujarat, Maharashtra, Goa, Karnataka and Kerala.

4.6.6 Landslides

- The term' landslide' includes all varieties of mass movements of hill slopes and can be defined as the downward and outward movement of slope forming materials composed of rocks, soils, artificial fills or combination of all these materials along surfaces of separation by falling, sliding and flowing, either slowly or quickly from one place to another.
- Although the landslides are primarily associated with mountainous terrains, these can also occur in areas where an activity such as surface excavations for highways, buildings and open pit mines takes place.
- They often take place in conjunction with earthquakes, floods and volcanoes.

- At times, prolonged rainfall causing landslide may block the flow of river for quite some time. The formation of river blocks can cause havoc to the settlements downstream on its bursting.
- Mainly affect the Himalayan region and the western ghats of India. Landslides are also common in the Nilgiri range.
- It is estimated that 30 percent of the world's landslides occur in the Himalayas. The Himalayan Mountains, which constitute the youngest and most dominating mountain system in the world, are not a single long landmass but comprises a series of seven curvilinear parallel folds running along a grand arc for a total of 3400 kilometers.
- Due to its unique nature, the Himalayas have a history of landslides that has no comparison with any other mountain range in the world. Landslides are also common in western ghat.
- In the Nilgiris, in 1978 alone, unprecedented rains in the region triggered about one hundred landslides which caused severe damage to communication lines, tea gardens and other cultivated crops. A valley in Nilgiris is called "Avalanches Valley".
- Scientific observation in north Sikkim and Garhwal regions in the Himalayas clearly reveal that there is an average of two landslides per sq. km. The mean rate of land loss is to the tune of 120 meter per km per year and annual soil loss is about 2500 tones per sq km.
- Landslides constitute a major natural hazard in our country, which accounts for considerable loss of life and damage to communication routes, human settlements, agricultural fields and forest lands. Based on the general experience with landslides, a rough estimate of monetary loss is of the order of 100 crore to 150 crore per annum at the current prices for the country as a whole.

Typical adverse effects:

- Damage occurs to human settlement, buildings, structures and infrastructure, especially bridges, elevated roads, railways, water towers, pipelines, electrical generating facilities. Aftershocks of an earthquake can cause much greater damage to already weakened structures.
- Secondary effects include fires, dam failure and landslides which may block water ways and also cause flooding.
- Damage may occur to facilities using or manufacturing dangerous materials resulting in possible chemical spills.
- There may also be a breakdown of communication facilities.
- The effect of an earthquake is diverse.
- There are large number of casualties because of the poor engineering design of the buildings and close proximity of the people.
- About 95 per cent of the people who are killed or who are affected by the earthquake is because of the building collapse.
- There is also a huge loss to the public health system, transport and communication and water supply in the affected areas.

4.7 SITE SAFETY

4.7.1 Necessity

- More workers meet with accidents during the 1st year of their employment.
- There were more accidents caused by falling (working at height), persons being struck or trapped by objects in motion, scaffolds, electric shocks, construction machinery etc.
- The possibility of labor accidents arising in the construction industry as opposed to all industries is over 40% and the percentage of fatalities is also exceedingly high. Therefore, safety measures to be employed in the construction industry continue to be matters of urgent importance that must be further enforced.
- Construction sites in developing countries are 10 times more dangerous than in industrialized nation and Indian sites are no exception.
- An analysis of occupational accidents reveals that the accidents caused due to highly unsafe work conditions and actions account for 80% of all accidents.
- In absence of "safety culture" in industry (unlike manufacturing sector) accident injuries and deaths always remained on the top (next to road accidents) at the national level.

- In order to reduce potential occupational accidents and cut back occupational injuries, it is necessary that each workplace should adopt a management system for occupational safety and health, precisely identifying risk factors under this system and systematically and continuously develop the measures that should be taken to eliminate or reduce those risk factors.
- It is important to continuously strive to reduce latent risk factors at workplaces as close to zero as possible.
- Liberalization, privatization and globalization of Indian economy and consequent foreign investments in infra structural sector have changed the construction scenario dramatically. Multinationals have brought with them high standards of safety and health stipulations which Indian construction industry was not used to. In order to compete with these multinationals, it is essential that the Indian construction companies ensure that the safety management function is integrated with overall project management system and safety culture is inculcated through, occupational safety health and environment management system.
- It is necessary for the central government, employers, and workers and people in general to aggressively grapple with adherence to safety by them and strive to enhance the safety awareness of society as a whole. 10. It would be important to create safety culture as an issue common to the nation as a whole or, in other word, the organizations and individuals to foster their mindset in which safety takes top priority.

Preventing construction accidents

Major construction accidents at site were prevented by taking proactive actions such as:

- Heavy equipment accidents were prevented by;
  - developing operating procedures and their implementation.
  - employing trained and skilled persons to operate.
  - providing adequate signaling and supervision.
  - providing overload preventing devices.
  - installing warning devices such as back sensors, contact sensors, flashers equipped with warning buzzers.
  - periodic preventive maintenance and upkeep.
- Landslide accidents, collapse of natural ground were prevented by maintaining a safe excavation grade and height and by constructing protective shoring. Falling accidents were prevented by making work floors and pathways as flat and slip proof as possible and building handrails, fences in the work area, obstructing openings with icaffolding safety nets, safety ropes etc.
- Providing nets, sheeting, reinforcement using bracing cable or other means prevented collapses of scaffolding/ structures.
- Accidents caused by falling objects during construction work were prevented by construction plan that avoids work being performed in both high and low work areas at the same time by using netting, construction sheeting, mesh sheeting, by binding materials together when placing them in high work areas, by building a drop chute and prohibiting access to hazardous area.
- All workers were given appropriate personal protective equipments (P.P.E.) while performing their jobs in order to protect their body parts from variety of hazards. Strict enforcement to use P.P.E. (by both men and women) and maintenance of P.P.E. resulted in significant reduction in minor injuries as well as reduction in number of first aid services.

Safety in construction industry

- The construction industry is one of the oldest and most hazardous industries contributing to a high rate of fatalities and serious accidents. Statutory protection has now been provided through Construction Safety Act or otherwise known as "Building and other Construction Workers (Regulation of Employment and Conditions of Services) Act, 1996.
- The Act has been made to provide Safety, Health and Welfare measures and give construction workers a Social Security.

4.7.2 Principles

- Principle 1: Demonstrate Safety Leadership
- Principle 2: Promote design for safety

CORE ELEMENTS OF THE RECOMMENDED PRACTICES FOR SAFETY AND HEALTHY PROGRAMS IN CONSTRUCTION

MANAGEMENT LEADERSHIP	• Top management demonstrates its commitment to eliminating hazards and to continuously improving workplace safety and health, communicates that commitment to workers and sets program expectations and responsibilities. • Managers at all levels make safety and heath a core organizational value, establish safety and health goals and objectives, provide adequate resources and support for the program and set a good example.
WORKERS PARTICIPATION	• Workers and their representatives are involved in all aspects of the program-including setting goals, identifying and reporting hazards, investigating incidents and tracking progress. • All workers, including contractors and temporary workers, understand their roles and responsibilities under the program and what they need to do to effectively carry them out. • Workers are encouraged and have means to communicate openly with management and to report safety and health concerns of suggest improvement, without fear of retaliation. • Any potential barriers or obstacles to workers participation in the program (for example, language, lack of information or disincentives) are removed or addressed.
HAZARD IDENTIFICATION AND ASSESSMENT	• Procedures are put in place to continually identify workpiece hazards and evaluate risks. • Safety and health hazards from routine, non-routine and emergency situation are identified and assessed. • An initial assessment of existing hazards, exposures and control measures in followed by periodic inspections and reassessments to identify new hazards. • Any incidents are investigated with the goal if identifying the root causes. • Identified hazards are prioritized for control.
HAZARD PREVENTION AND CONTROL	• Employers and workers co-operate to identify and select methods for eliminating preventing or controlling workplace hazards. • Controls are selected according to a hiearachy that uses engineering solutions first, followed by safe work practices, administrative controls and finally personal protective equipment (PPE). • A plan is developed that ensures controls are implemented, interim protection is provided progress is tracked and the effectiveness of control is verified.
EDUCATION AND TRAINING	• All workers are trained to understand how the program works and how to carry out the responsibilities assigned to them under the program. • Employers, managers and supervisors receive training on safety concepts and their responsibility for protecting worker's rights and responding to workers' reports and concerns. • All workers are trained to recognize workplace hazards and to understand the control measures that have been implemented.
PROGRAM EVALUATION AND IMPROVEMENT	• Control measures are periodically evaluated for effectiveness. • Processes are established to monitor program performance, verify program implementation, and identify program shortcomings and opportunities for improvement. • Necessity action are taken to improve the program and overall safety and health performance.
COMMUNICATION AND CO-ORDINATION FOR EMPLOYERS ON MULTIEMPLOYER WORKSITES	• General contactors, contractors and staffing agencies commit to providing the same level of safety and health protection to all employees. • General contractors, contractors, subcontractors and staffing agencies communicate the hazards present at the worksite and the hazards that work of contract workers may create on site. • General contractors establish specification and qualification for contractors and staffing agencies. • Prior to bringing work, general contractors, contractors and staffing agencies co-ordinate on work planning and scheduling to identify and resolve any conflicts that could impact safety or health.

4.7.3 Tools, techniques

SAFETY MEASURES DURING CONSTRUCTION

- **Safety management in general:** Specific responsibility by name to be allotted at a specific point of time.
- **Planning of housekeeping:** The construction site must always be kept in neat and tidy condition. All materials must be stacked in proper place and in proper manner with proper marking.
- **A well planned site helps reduce accidents.**
- **Safety while excavating**:
  1. Provide side supports in an excavated trench.
  2. Always wear a safety helmet when you work in an excavation.
  3. Hand dig with care, as cables may be just below the surface.
- **Scaffolding:** Make sure that scaffolds are securely anchored or tied to the building.
- **Ladders:**
  1. Make sure your ladder is lashed or footed before you climb it.
  2. Avoid carrying tools or materials in your hand while you are climbing ladders.
- Timber used in scaffolding and ladders should not be painted.
- **Movement of materials:**
  1. While lifting the load through crane use a signaler and see that load is fitted with a safety hook.
  2. Never climb-up the tower or get into the boom while the crane is in use.
  3. Never ride on the platform of a goods hoist.
  4. No one should stand within the operating area of the crane while the load is in motion.
- **Working Positions, tools and equipments**:
  1. Arrange to sit whenever possible.
  2. Keep materials, tools and controls within easy reach.
  3. Carry tools in tool holders and not in the pockets of clothing.
  4. A dangerous part of machinery requires guard – a warning notice is not a substitute.
- **Working environment:**
  1. Never use solvents to remove paint or grease from your skin.
  2. The dust you cannot see is more dangerous than the dust you can see.
- **Personal protective equipment (PPE):**
  1. Your safety helmet protects you only if you have it on. Ninety percent of all eye injuries can be prevented by Suitable eye protection.
  2. Make a habit of using Personal Protective Equipment.

1. **Safety Manual:** Prepare a safety manual for company.
2. **Safety Co-ordinator:** Depute full time/part time safety co-ordinators depending upon the size of the project site and nature of hazards.
3. **Worker Screening**: Screening of workmen.
4. **Project Safety Plan:** Prepare a Project Safety PI = and implement at job sites which covers:
   (a) Overall scope of work: Fabrication, erection, excavation, concreting, brickwork, finishing work, painting, blasting, non-distructive tests etc.
   (b) Obligation chart: Responsibility of the line functionaries and reporting of the safety engineering functions.

(c) Safety committee (all job sites):
- Structure: Chairman, Secretary and Members
- Roles and responsibilities
- Project Manager
- Safety committee members
- Safety coordinators
- Site engineers
- First line supervisors/Engineers
- Subcontractors.

(d) Job related specific P.P.E. (personal protective equipments)

(e) General site safety rules:
- Use of P.P.E. and safety devices relevant to site activities.
- General housekeeping, stacking of materials.
- Road safety inside the projects.
- Access/egress and workstation safety.
- Safe use of construction power.
- **Work permit system:**
  - Hot work permit.
  - Excavation clearance permit.
  - Industrial radiography permit.
  - Permit to work on overhead traction line.
  - Permit to work on plant and machinery and other power driven equipment.
  - Electrical work permit.
  - Permit to open manhole covers/grills.
  - Permit for confined space entry.

(f) Use, maintenance and inspection of plant and machinery.

(g) Scaffolding norms.

(h) Use of safe materials handling and personal handling equipment.

(i) Fire prevention and protection.

(j) Contingency plan.

(k) Others.

5. Identification of High Risk/Hazardous Operations in the Site and their Appropriate Safety Measures:

(a) Structural and equipment erection.

(b) Excavation.

(c) Concreting - Chimney, silos, cooling towers etc.

(d) Pre-casting and pre-cast erection.

(e) Pile driving.

(f) Road works: existing roads/new roads.

(g) Work over water.

(h) Blasting.

(i) Use, handling, storage of compressed gas cylinders etc. (j) Pressure testing — hydraulic and pneumatic.

(k) Tunneling. (1) Others.

6. **Peptalk.**
7. **Checklists and Reports**

   General site safety inspections: Excavation, demolition, blasting, scaffoldings, ladders, working plat-forms, PPE, structural fabrication and erection, gas cutting and welding, means of access, electrical work, housekeeping, fire prevention and protection, road safety, first aid box, illumination, safety boards, post-ers, stickers, drinking water, sanitation, emergency vehicles.

   - Electrical safety inspection.
   - Erection safety inspection.
   - Scaffolding inspection.
   - Safety while working at height.
   - Cranes inspection.
   - Vehicle and earth moving equipment inspection.
8. **Statistics and Reports**
9. **Accident Investigation Report**
   (a) Action taken report.
   (b) Weekly analysis of accidents.
   (c) Monthly safety statistics.
   (d) Register of first aid.
   (e) Subcontractors' safety performance evaluation.
10. **Job Safety Analysis**

    Identity of task: Asbestos roof sheeting at 25 meters height.
11. **Third Party (competent person) Inspection, examination and testing of :**
    - Hoists and Lifts, once in six months.
    - Lifting machines (crane, crab, winch, pulley block etc.) and lifting tackles (chain, rope sling, hook, D-shackle, coupling, spider or similar appliances, once in 12 months.
12. **Safety Training**
13. **Safety Awareness Programme.**

4.7.4 Laws, rules and regulations

The general safety rules for construction workers

- Always be mindful on your job. No horseplay should be done while on job.
- Always wear all the personal protective equipment for the job.
- Protect and prevent yourself from hot materials.
- In case of contact with acid like substances immediately wash it with plenty of water.
- Behave properly, walk properly; do not run in construction site.
- The use of drugs and alcohol is strictly prohibited in construction site as these are the causes of many serious accidents. If you use some strong medicine with the advice of doctors, then it should be known to your immediate supervisor, so that he can allot you the jobs accordingly.
- Do not try to attract the attention of others while doing the job.
- Sanitation facilities are provided for you and for your colleagues. Try to keep it clean as far as it is possible. Do not try to deface or degrade or damage these facilities. These are forbidden acts and try to keep away from such jobs.
- Cleanliness is the next to Godliness also clean job is the start of a safe job. Develop the habit of good housekeeping.
- Compressor air is the most costly form of energy and it also contains moisture and dirt etc. So do not use compressed air for the removal of dust from your clothes and from your body. This may seriously harm you.

- Do not work at height if you are afraid to do so, also if you feel dizziness, nausea, and vomiting tendency then it is better to avoid working at height.
- Keep the injured as comfortable as possible. Never move the injured person unless it is essential. Utilize the first-aid kit available at the site so long the ambulance arrives.
- It is your duty to know the location of nearest fire point from your place of work. You never know at what moment you need it. Also get yourself trained how to use different fire extinguishers.
- Know the correct method of manual lifting. Lift correctly with legs, not by the back. If load is too heavy for a single man to lift, take help from others. Stay fit and controls your weight. It is a good practice to start the job after doing some stretching exercises. It has been observed that nearly 20% of injuries occurred during lifting of materials.
- Only the authorized drivers are allowed to drive all heavy earth moving machineries. No body other than driver should be allowed inside the driver's cabin.
- Do not use power tools and equipments until you have been properly instructed to do so and authorized to use them.
- Ensure that all machinery guards are in its place. Do not remove, displace, damage, or destroy any safety device provided on the machinery.
- Do not enter inside a barricaded area.
- If you must work around power shovels, trucks, and dozers, make sure that operators can always see you. Barricades are required for cranes.
- Never lubricate a machine in running condition.
- Disconnect the power source, lock out, and properly tag out the power tools before servicing, repairing or adjusting.
- Barricade all the danger areas.
- Excavations over1.5 m deep must be shored or sloped as required. Keep yourself out of trenches or cuts that have not been properly shored or sloped. Excavated or other materials shall not be stored nearer than 1.5 m from the edge of the excavation.
- Use the 'four and one' rule when using a ladder, i.e., one foot of base for every four feet of height. feet unless ladder is tied, blocked or otherwise secured. safety
- Portable ladders in use shall be equipped with Step ladders shall not be used as a straight ladder.
- Ladders must extend three feet above landing on roof for proper use.
- Defective ladders must be properly tagged and removed from service.
- Keep ladder bases free of debris, hoses, wires, materials etc.
- Build scaffolds according to the manufacturers' recommendations and construction safety standards.
- Scaffold planks shall be properly lapped, cleated or otherwise secured to prevent shifting.
- Use extension cords of three-prong type only. Use ground fault circuit interrupters at all times and when using tools in wet conditions; check the electrical grounding systems daily.
- The use of safety harness in height work is essential. Use your safety lines anchored properly.
- Never throw anything 'overboard'. Someone passing below may be seriously injured.
- Open fires are prohibited in construction site.
- Know what emergency procedure is established at the site you work like location of emergency phone numbers, location of first-aid kit, location of stretcher, location of fire points and evacuation plans etc.
- Never enter a manhole, well, shaft, tunnel, or other confined space which could possibly have a non respirable atmosphere because of lack of oxygen or presence of toxic or flammable gas. Make certain a competent person tests the confined area with an appropriate detector before entry and necessary safety equipments must be worn. Stand-by persons should be stationed at the entrance.

Construction safety act

- The construction industry is one of the oldest and most hazardous industries contributing to a high rate of fatal and serious accidents. Statutory protection has been provided to the building and construction workers through this Act.
- The construction safety act is called as "Building and other construction workers (Regulation of Employment and Conditions of Services), Act, 1996" (No. 27 of 1996) has been passed by the parliament and received the assent of the President on 19th August, 1996. The Act is applicable with effect from 1st March 1996.
- To regulate the employment and conditions of service of building and other construction workers and to provide for their safety, health and welfare measures and for other matters connected therewith or incidental thereto.
- Every establishment which employs, or had employed on any day of the preceding 12 months, 50 or more building workers in any building or other construction work.
- **Building and other construction work:** The construction, alteration, repairs, maintenance or demolition of, or in relation to buildings, streets, roads, rail-ways, train ways, air fields, irrigation, drainage, embankment and navigation works, flood control works, generation, transmission and distribution of power, water works, oil and gas installations, electric lines, wireless, radio, television, telephone, telegraph and overseas communications, dams, canals, reservoirs, water courses, tunnels, bridges, viaducts, aqua ducts, pipe lines, towers, cooling towers, transmission towers etc. but does not include any building of other construction work to which the provisions of the Factories Act, 1948 or the Mines Act, 1952 apply.
- **Building worker:** A person who is employed to do any skilled, semiskilled or u7.-skilled, manual, supervisory, technical or clerical world

Appropriate Government:

- In relation to an establishment (which employs building workers either directly or through a contractor) in respect of which the appropriate Government under the Industrial Disputes Act, 1947 (14 of 1947), is the Central Government, the Central Government.
- In relation to any such establishment, being a public sector undertaking, as the central government may by notification specify which employs building workers either directly or through a contractor, the Central Government.

Main Provisions Relating to Safety

- **Constitution of Safety Committee:** The employer of every establishment ordinarily employing five hundred or more building workers shall constitute a safety committee of such number of representatives of the employer and equal number of the building workers.
- **Appointment of Safety Officer:** The employer of every establishment ordinarily employing five hundred or more building workers shall appoint a Safety Officer possessing a professional qualification.
- **Notice of Certain Accidents:** Employer of any establishment shall give notice of an accident which causes death or which causes bodily injury because of which the person injured is prevented from the working for a period of 48 hours or more following the accident. The concerned authority would make investigation if accident causes death of five or more persons. It shall make an inquiry into such accident within one month of receipt of the notice.
- **Registration of Establishment:** All Assistant Labor Commissioners (Central) have been notified as Registering Officer under the Act. Every establishment shall make an application to the Registering Officer for the registration of establishment within 'a period of 60 days from the commencement of the Act. The Registering Officer shall register the establishment and issue a certificate of registration to the employer.
- **Responsibility of Employer:** An employer shall be responsible for providing constant and adequate supervision of any building or other construction work in the establishment to ensure compliance with the provisions of this Act relating to safety and for taking practical steps necessary to prevent accidents.

- **Rules for the Safety and Health of Building Workers:** The appropriate Government may makes rules regarding the measures to be taken for the safety and health of the building workers in the course of their employer and the equipments and appliances necessary to be provided to them for ensuring their safety, health and protection.

  Such rules may provide for all or any of the following matters:

- The safety means of access to, and the safety of, any working place, including the provision of suit-able and sufficient scaffolding at and various stages when work cannot he done safely from the ground or from any part of a building or from a ladder or such other means of support.
- The precautions to be taken in connection with the demolition of the whole or any substantial part of a building or other structure under the supervision of a competent person and the avoidance of the danger from collapse of any building or other structure while removing any part of the framed building or other structure by shoring or otherwise.
- The handling and use of explosive under the control of competent person so that there is no exposure to the risk of injury from explosion or from flying material.
- The erection, installation, use and maintenance of transporting equipment, such as locomotives, trucks, wagons and other vehicles and trailers and appointment of competent person to drive or operate such equipment.
- The erection, installation, use and maintenance of hoists, lifting appliances and lifting gear including periodical testing and examination and heat treatment, where necessary, precautions to be taken while raising or lowering loads, restrictions on carriage of persons and appointment of competent persons on hoists or other lifting appliances.
- The adequate and suitable lighting of every work-place and approach there to, of every place where raising or lowering operations with the use of hoists, lifting appliances or lifting gears are in progress and of all operations dangerous to build-ing workers employed.
- The precautions to be taken to prevent inhalation of dust, fumes, gases or vapours during any grinding, cleaning, spraying or manipulation of any material and steps to be taken to secure and maintain adequate ventilation of every working place or confined space.
- The measures to be taken during stacking and un-stacking, stowing or unstowing of materials or goods or handling in connection there with.
- The safe guarding of machinery including the fencing of every fly-wheel and every moving part of a prime mover and every part of transmission or other machinery, unless it is in such a position or of such construction as to be safe to every worker working on any of the operations and as if it were securely fenced.
- The safe handling and use of plant, including tools and equipment operated by compressed air.
- The precautions to be taken in case of fire.
- The limits of weight to be lifted or moved by workers.
- The safe transport of workers to or from any work-place by water and provision of means for rescue from drowning.
- The steps to be taken to prevent danger to workers from live electric wires or apparatus including electrical machinery and tools and from electrical wires.
- The keeping of safety nets, safety sheets and safety belts where the special nature or the circumstances of work render them necessary for the safety of the workers.
- The standards to be complied with regard to scaffolding, ladders and stairs, lifting appliances, ropes, chains and accessories, earth moving equipment and floating operational equipment.
- The precaution to be taken with regard to pile driving, concrete work, work with hot asphalt, tar or other similar things, insulation work, demolition operations, excavation, underground construction and handling materials.

- The safety policy, that is to say, a policy relating to steps to be taken to ensure the safety and health of building workers, the administrative arrangements therefore and the matter connected therewith, to be framed by the employees and contractors for the operations to be carried on in a building or other construction work.
- The information to be furnished to the Bureau of Indian Standards established under the Bureau of Indian Standards Act, 1986 (63 of 1986), regarding the use of any article or process covered under the Act in a building or other construction work. The provision and maintenance of medical facilities for building workers.
- The Central Government may frame model rules in the matters as mentioned above. The rules in this respect are being framed and will be notified.

Inspecting staff

Director General of Inspection

- The Central Government has appointed the Chief Labor Commissioner (Central) as the Director General of Inspectors under the Act. He is responsible for laying down the standards of inspection and also exercise the powers of an Inspector throughout India in relation to all the establishments for which the Central Government is the appropriate government.

Chief Inspector of Inspection

- The State Government may appoint Chief Inspector of Inspection of Building and Construction responsible for effectively carrying out the provisions of this Act in the State and shall also exercise powers of an Inspector throughout the state in relation to establishments for which the State Government is the appropriate Government.

Inspectors

- The Government of India has appointed the following officers to be the Inspectors under the Act.
    - Joint Chief Labour Commissioner (Central) New Delhi.
    - All Deputy Chief Labor Commissioners (Central).
    - All Regional Labour Commissioners (Central).
    - All Assistant Labour Commissioners (Central).
    - Chief Advisory (Labour Welfare) Office (Central), New Delhi.
    - All Labour Enforcement Officers (Central).

Powers of Inspectors

(a) Enter, at all reasonable hours, with such assistants (if any) being persons in the service of the Government or any local or other public authority as he thinks fit, any premises or place where building or other construction work is carried on, for the purpose of examining in any register or record or notices required to be kept or exhibited by or under this Act, and require the production thereof for inspection.

(b) Examine any person whom he finds in any such premises or place and who, he has reasonable cause to believe, is a building worker employed there in.

(c) Require any person giving out building or other construction work to any building worker, to give any information, which is in his power to give with respect to the names and addresses of the persons to, for and whom the building and other construction work is given out or received, and with respect to the payments to be made for the building and other construction work.

(d) Seize or take copies of such register, record of wages or notices or portions thereof as he may consider relevant in respect of an offence under this Act which he has reason to believe has been committed by the employer.

Other Applicable Statutes

- Apart from the legislations entitled "Building and Other Construction Workers (Regulation of Employment and Conditions of Services) Act 1996." Following legislations are also applicable:
  - Mines Act — 1952 — quarries, underground work.
  - Motor Vehicles Act — Vehicles.
  - Shops and Establishment Act — Offices.
  - Contract Labor Act — Contract labour.
  - Interstate Migrant Workers Act — Immigrant labour.
  - Indian Explosives Act — explosives.
  - Indian Electricity Act — use of Electricity.
  - Workmen's Compensation Act — Compensations.
  - Regulations Against Use of Child Labor.
  - The Petroleum Act 1934 and Rules 1976.
  - The Gas Cylinder Rules 1981.
  - The Static and Mobile Pressure Vessels (Unified) Rules 181. Feulation of Radiography.
  - Air and Water Pollution Control Regulations.
- Since construction industry is rapidly developing so all the rules and regulations must be known to the authorities for systematic completion of the project.

Important Points

- Water resources management (IWRM) has been defined by the Global Water Partnership (GWP) as and quota process which promotes the co-ordinated development and management of water, land and related resources, in order to maximize the resultant economic and social welfare.
- Many communities accept that there is growing demand for water and thinking ahead about water security or management. Methods of water management can be classified as Conservation, Allocation, retrofit program and Behavioural practices.
- Water Resources Management is a process which promotes the coordinated development and management of water, land and related resources, in order to maximise the resultant economic and social welfare in an equitable manner without compromising the sustainability of vital ecosystems.
- Top 10 Water Management Techniques

  Meter/Measure/Manage.

  Optimize Cooling Towers.

  Replace Restroom Fixtures.

  Eliminate Single-Pass Cooling.

  Use Water-Smart Landscaping and Irrigation.

  Reduce Steam Sterilizer Tempering Water Use.

  Reuse Laboratory Culture Water.

  Control Reverse Osmosis System Operation.
- The waste management hierarchy breaks this process down into four parts, also known as the Four R (Refuse, Reduce, Reuse, and Recycle).
- RECOVER is to convert waste into resources (such as electricity, heat, compost and fuel) through thermal and biological means.
- **Reduce :** Reducing the amount of products/ recourse we consume. Reducing what we consume reduces the amount of garbage that litters the land and occupies space in community landfills.
- "Zero Waste is a goal that is ethical, economical, efficient and visionary, to guide people in changing their lifestyles and practices to emulate sustainable natural cycles, where all discarded materials are designed to become resources for others to use.

- There is very good potential to recycle many elements of construction waste. Recycling involves collecting, reprocessing and/ or recovering certain waste materials to make new materials or products. Often roll-off containers are used to transport the waste. Rubble can be crushed and reused in construction projects.
- Reuse/recycling from deconstruction/demolition sites.

  Site works and vegetation – asphalt paving, chain link fencing, timber fencing, trees.

  Concrete – in situ and precast concrete.

  Masonry – concrete blocks and decorative concrete, paving stones, bricks.
- Fly ash is a fine powder that is a byproduct of burning pulverized coal in electric generation power plants. ... When used in concrete mixes, fly ash improves the strength and segregation of the concrete and makes it easier to pump.
- Fly ash use in concrete improves the workability of plastic concrete, and the strength and durability of hardened concrete. Fly ash use is also cost effective. When fly ash is added to concrete, the amount of portland cement may be reduced. Benefits to Fresh Concrete.
- Air entrainment agent is used to toughen the bricks. Since the concentration of calcium oxide is very high in class C fly ash, the brick is described as self-cementing.
- Granulated blast furnace slag has been used as a raw material for cement production and as an aggregate and insulating material. ... Ground granulated blast furnace slag is used commercially as a supplementary cementitious material in Portland cement concrete (as a mineral admixture or component of blended cement).
- **Principal uses include:** Blending with many other products such as granulated slag, fly ash and lime to form pavement material. Other uses Include, skid resistant asphalt aggregate, rail ballast asphaltic concrete aggregate, soil conditioner, hard stand areas and unconfined construction fill.
- Plastic roads mainly use plastic carry bags, disposable cups and PET bottles that are collected from garbage dumps as important ingredients of the construction materials. By using plastic waste as modifier, we can reduce the quantity of cement and sand by their weight, hence decreasing the overall cost of construction.
- Types of Renewable Energy
  1. Wind Power. The U.S. can produce more than 10 times its energy needs through wind power, one of the fastest-growing energy sources.
  2. Solar Power. The sun will be a source of energy for billions of years.
  3. Biomass. This is the use of plant matter and animal waste to create electricity.
  4. Geothermal Energy.
  5. Hydroelectric Power.
- Solar cells can be used to generate electricity from sunlight. It is a device that converts light energy into electrical energy. Sometimes the term solar cell is reserved for devices intended specifically to capture energy from sunlight, while the term photovoltaic cell is used when the light source is unspecified.
- A few of the lesser-known benefits of solar energy Zero energy-production costs.

  Less energy lost during long-distance transport.

  Versatile installation.

  Energy production coincides with the times of maximum demand.

  No Impact on the environment.

  Improved security of the power grid.

  Economic savings.

  Good Economic impact.

- Biomass is organic material that comes from plants and animals, and it is a renewable source of energy. ... When biomass is burned, the chemical energy in biomass is released as heat. Biomass can be burned directly or converted to liquid biofuels or biogas that can be burned as fuels.
- We use four types of biomass today—wood and agricultural products, solid waste, landfill gas and biogas, and alcohol fuels (like Ethanol or Biodiesel). Most biomass used today is home grown energy. Wood—logs, chips, bark, and sawdust—accounts for about 44 percent of biomass energy.
- Biomass is the only renewable energy source that can be converted into liquid biofuels such as ethanol and biodiesel. Biofuel is used to power vehicles, and is being produced by gasification in countries such as Sweden, Austria, and the United States.
- Biomass used as a fuel reduces need for fossil fuels for the production of heat, steam, and electricity for residential, industrial and agricultural use. Biomass is always available and can be produced as a renewable resource.
- Wind energy is a form of solar energy. Wind energy (or wind power) describes the process by which wind is used to generate electricity. Wind turbines convert the kinetic energy in the wind into mechanical power. ... Mechanical power can also be utilized directly for specific tasks such as pumping water.
- Wind turbines convert the kinetic energy in the wind into mechanical power. This mechanical power can be used for specific tasks (such as grinding grain or pumping water) or a generator can convert this mechanical power into electricity to power homes, businesses, schools, and the like.
- Advantages of Wind Power

  Wind power is cost-effective.

  Wind creates jobs.

  Wind enables industry growth and competitiveness.

  It is a clean fuel source.

  Wind is a domestic source of energy.
- It is sustainable.

  Wind turbines can be built on existing farms or ranches.
- Disadvantages of Wind Energy.

  The Wind Fluctuates. Wind energy has a similar drawback to solar energy in that it is not constant.

  Installation is Expensive. Although costs are reducing over time, wind turbines are still expensive.

  Wind Turbines Pose a Threat to Wildlife.

  Wind Turbines Create Noise Pollution.

  They Also Create Visual Pollution.
- Waves are caused by the wind blowing over the surface of the ocean.
- Ocean waves contain tremendous energy potential. Wave power devices extract energy from the surface motion of ocean waves or from pressure fluctuations below the surface.
- There are three basic types that allow us to use ocean for its energy. We can use the waves (wave energy, wave power), ocean tidal power (ocean high and low tides), and we can even use temperature differences in the water to create an energy (Ocean Thermal Energy Conversion, OTEC).
- Wave energy is produced when electricity generators are placed on the surface of the ocean. The energy provided is most often used in desalination plants, power plants and water pumps. Energy output is determined by wave height, wave speed, wavelength, and water density.

- Basic principle of tidal power plant A dam is constructed in such a way that a basin gets separated from the sea and a difference in the water level is obtained between the basin and sea. The constructed basin is filled during high tide and emptied during low tide passing through sluices and turbine respectively.
- Geothermal energy is heat derived within the sub-surface of the earth. Water and/or steam carry the geothermal energy to the Earth's surface. Depending on its characteristics, geothermal energy can be used for heating and cooling purposes or be harnessed to generate clean electricity.
- People can capture geothermal energy through: Geothermal power plants, which use heat from deep inside the Earth to generate steam to make electricity. Geothermal heat pumps, which tap into heat close to the Earth's surface to heat water or provide heat for buildings.
- Geothermal Energy Disadvantages

  Potential emissions – Greenhouse gas below Earth's surface can potentially migrate to the surface and into the atmosphere.

  Surface Instability – Construction of geothermal power plants can affect the stability of land.
- Hybrid power system. Hybrid systems, as the name implies, combine two or more modes of electricity generation together, usually using renewable technologies such as solar photovoltaic (PV) and wind turbines.
- Hybrid power plants often contain a renewable energy component (such as PV) that is balanced via a second form of generation or storage such as a diesel genset, fuel cell or battery storage system. They can also provide other forms of power such as heat for some applications.
- Energy audit is an inspection survey and an analysis of energy flows for energy conservation in a building. It may include a process or system to reduce the amount of energy input into the system without negatively affecting the output.
- An energy audit is an assessment of your home that takes a look at current energy consumption and then identifies energy efficiency measures that you can conduct to make your home more efficient.
- An office energy audit checklist is used to assess the energy efficiency of a workspace based on equipment, appliances, design, and usage. Certified energy auditors can use this checklist to identify opportunities for energy cost reduction and recommend solutions.
- A home energy audit is a process that helps you to identify where your home is losing energy and what steps you can take to improve energy efficiency. A home energy audit is something that every homeowner should be doing every two to three years in order to ensure maximum efficiency and spending.
- A natural disaster is a major adverse event resulting from natural processes of the Earth; examples are floods, hurricanes, tornadoes, volcanic eruptions, earthquakes, tsunamis, and other geologic processes.
- Types of Disaster

  Geophysical (for example, Earthquakes, Landslides, Tsunamis and Volcanic Activity).

  Hydrological (for example, Avalanches and Floods).

  Climatological (for example, Extreme Temperatures, Drought and Wildfires).

  Meteorological (for example, Cyclones and Storms/Wave Surges).

  Biological (for example, Disease Epidemics and Insect/Animal Plagues).
- The leading safety hazards on construction sites include falls, being caught between objects, electrocutions, and being struck by objects. Falls are the leading cause of injury in the construction industry, in particularly for elder and untrained construction workers.
- Health and safety in construction are particularly important because the industry is prone to hazardous situations and can be dangerous at times. There are some major reasons why health and safety is important in the workplace.

- Equipment and material theft from construction job sites is a major source of loss for contractors. A source of some theft is unauthorized intruders, but unauthorized people gaining access to the job site also result in safety and injury issues, liability issues, and the potential equipment damage.

Multiple Choice Questions

1. A co-ordinated development and management of water is called ......
   (a) water resource management. (b) rain water management.
   (c) water supply management. (d) water saving management.
2. Recover in waste management means ......
   (a) convert waste in to a compost. (b) convert waste in to an energy.
   (c) convert waste in to reusable material. (d) all of these above.
3. Demolition waste obtained fro construction sites is used for ......
   (a) construction of walls. (b) filling of trenches.
   (c) construction of slab (R.C.C.). (d) concrete making.
4. Fly ash used for construction in concrete to ......
   (a) increase strength. (b) decrease segregation.
   (c) increase workability. (d) all of these.
5. Blast furnace slab is obtained from ......
   (a) under ground excavations. (b) steel plants.
   (c) vegetations in jungle. (d) glass.
6. Plastic roads constructed from the plastic waste are ......
   (a) less durable. (b) economical.
   (c) not suitable for our country. (d) none of these.
7. The cell which converts energy from sun into electric energy is ......
   (a) denial cell. (b) secondary cell.
   (c) button cell. (d) solar cell.
8. Biomass is obtained from ......
   (a) ore. (b) plants.
   (c) animals. (d) plants and animals.
9. A machine which converts wind energy into mechanical energy is called ......
   (a) wind turbine. (b) wind motor.
   (c) wind pump. (d) none of these.
10. A machine which converts mechanical energy into electrical energy is called ......
    (a) a pump. (b) an engine.
    (c) a generator. (d) a turbine.
11. "OTEC" means ......
    (a) ocean temperature and environmental concern.
    (b) ocean technology of environmental commission.
    (c) ocean thermal energy conversion.
    (d) ocean thermal engineering concept.
12. Geothermal energy is obtained from ......
    (a) jungle. (b) wind.
    (c) subsurface of earth. (d) sky.

13. Hybrid power system is a combination of ......
    (a) thermal energy and solar energy
    (b) thermal energy and wind energy
    (c) solar energy and wind energy
    (d) all of these
14. An inspection survey and analysis of flow of energy in a building is called ......
    (a) energy survey
    (b) energy inspection
    (c) energy saving
    (d) energy audit
15. Floods, frurricanes, volcanoes are types of ......
    (a) man made disaster
    (b) natural disaster
    (c) artificial disaster
    (d) by product disaster
16. Cyclones and storms are type of ......
    (a) geophysical disaster
    (b) climatological disaster
    (c) meteorological disaster
    (d) biological disaster
17. Various causes of accidents at a construction site are ......
    (a) falling of objects
    (b) improper safety
    (c) obstruction in movement
    (d) all of these
18. Earthquakes and landslides are types of ......
    (a) geophysical disaster
    (b) hydrological disaster
    (c) climatological disaster
    (d) biological disaster
19. Power generated from wind energy is ......
    (a) cost effective
    (b) very costly
    (c) requires fuel
    (d) none-sustainable
20. Solar energy is used to convert into electrical energy using ......
    (a) photovoltaic cell
    (b) ultra emission cell
    (c) X-rays cell
    (d) dry cell
21. How many blades does a modern wind turbine have ?
    (a) 3
    (b) 4
    (c) 2
    (d) none of these
22. Which nation leads the world in installed wind energy capacity ?
    (a) USA
    (b) India
    (c) China
    (d) Saudi Arab of Emirates
23. What is biomass energy ?
    (a) Energy generated from wind
    (b) Energy generated from plants and animals
    (c) Energy generated from river
    (d) Energy generated from nuclear power

Answers

1. (a)	2. (d)	3. (b)	4. (d)	5. (b)	6. (b)	7. (d)	8. (d)	9. (a)	10. (c)
11. (c)	12. (c)	13. (d)	14. (d)	15. (b)	16. (c)	17. (d)	18. (a)	19. (a)	20. (a)
21. (c)	22. (c)	23. (b)							

Chapter **5**

Advancement in Construction

Weightage of Marks = 18, Teaching Hours = 12

Syllabus

5.1 Building construction techniques

- Zero energy building, Green building, Mass housing-precast housing, prefab homes, pre-engineering building, Solar Paints, Building Integrated Photovoltaic (BIPV), Earthquake Resisting Controls-Isolation and Dissipation.

5.2 Road construction techniques

- 3D Printing, Road Printer, smart roads.

5.3 Coastal construction techniques

- Sound Proofing walls, water resistant roofs, high performance doors and windows, air and moisture barriers.

5.4 Ground improvement techniques

- Advanced piling techniques - Stone Column, Vibro Floatation, Micro Piles, Soil Nailing
- Vertical drains-Sand Drains, Pre-Fabricated Vertical Drains
- Thermal Methods- soil heating and soil freezing.

About this Chapter

At the end of this chapter, students will be able to:

- Select the appropriate advanced techniques for the given construction project.
- Understand the characteristic features of advanced construction techniques.
- Recommend the use of appropriate construction techniques as per site condition.
- Identify the application of given construction technique

5.1 BUILDING CONSTRUCTION TECHNIQUES

- Building construction methods have experienced significant facelift in recent times with innovative technologies being harnessed optimally for improving the qualitative index of buildings.
- This has spelled considerable advantages for end users like us who can remain immune from recurrent expenses on repairs and other incidental building-related jobs.
- Construction lead time has also been reduced and building costs have been rationalized.

5.1.1 Zero energy building

- A net zero-energy building (ZEB) is a residential or commercial building with greatly reduced energy needs through efficiency gains such that the balance of energy needs can be supplied with renewable technologies.
- A zero energy building can be defined in several ways, depending on the boundary and the metric. Different definitions may be appropriate, depending on the project goals and the values of the design team and building owner.

  For example, building owners typically care about energy costs. Organizations such as DOE are concerned with national energy numbers, and are typically interested in primary or source energy. A building designer may be interested in site energy use for energy code requirements. Finally, those who are concerned about pollution from power plants and the burning of fossil fuels may be interested in reducing emissions. Four commonly used definitions are: net zero site energy, net zero source energy, net zero energy costs, and net zero energy emissions.

- Most zero-energy buildings use the electrical grid for energy storage but some are independent of the grid and some include energy storage onsite.
- Energy is usually harvested onsite through energy producing technologies like solar and wind, while reducing the overall use of energy with highly efficient lightning and heating, ventilation and air conditioning (HVAC) technologies.
- The zero-energy goal is becoming more practical as the costs of alternative energy technologies decrease and the costs of traditional fossil fuels increase.
- The development of modern zero-energy buildings became possible largely through the progress made in new energy and construction technologies and techniques.
- These include highly insulating spray-foam insulation, high-efficiency solar panels, high-efficiency heat pumps and highly insulating, low emissivity, triple-glazed windows.

Design and construction

- The most cost-effective steps toward a reduction in a building's energy consumption usually occur during the design process.
- To achieve efficient energy use, zero energy design departs significantly from conventional construction practice.
- Successful zero energy building designers typically combine time tested passive solar, or artificial/fake conditioning, principles that work with the on-site assets.
- Zero-energy buildings are built with significant energy-saving features. The heating and cooling loads are lowered by using high-efficiency equipment (such as heat pumps rather than furnaces.
- Heat pumps are about four times as efficient as furnaces) added insulation (especially in the attic and in the basement of houses), high-efficiency windows (such as low emissivity, triple-glazed windows), draft-proofing, high efficiency appliances (particularly modern high-efficiency refrigerators), high-efficiency LED lighting, passive solar gain in winter and passive shading in the summer, natural ventilation, and other techniques.
- These features vary depending on climate zones in which the construction occurs. Water heating loads can be lowered by using water conservation fixtures, heat recovery units on waste water, and by using solar water heating, and high-efficiency water heating equipment.
- In addition, daylighting with skylights or solartubes can provide 100% of daytime illumination within the home.

Advantages

- Isolation for building owners from future energy price increases.
- Increased comfort due to more-uniform interior temperatures (this can be demonstrated with comparative isotherm maps).
- Reduced requirement for energy austerity.
- Reduced total cost of ownership due to improved energy efficiency.
- Reduced total net monthly cost of living.
- Reduced risk of loss from grid blackouts.
- Extra cost is minimized for new construction compared to an afterthought retrofit.
- Higher resale value as potential owners demand more ZEBs than available supply.
- The value of a ZEB building relative to similar conventional building should increase every time energy costs increase.

- Future legislative restrictions, and carbon emission taxes/penalties may force expensive retrofits to inefficient buildings.
- Contribute to the greater benefits of the society, for example, providing sustainable renewable energy to the grid, reducing the need of grid expansion.

Disadvantages

- Initial costs can be higher – effort required to understand, apply, and qualify for ZEB subsidies, if they exist.
- Very few designers or builders have the necessary skills or experience to build ZEBs.
- Possible declines in future utility company renewable energy costs may lessen the value of capital invested in energy efficiency
- New photovoltaic solar cells equipment technology price has been falling at roughly 17% per year – It will lessen the value of capital invested in a solar electric generating system – Current subsidies may be phased out as photovoltaic mass production lowers future price
- Challenge to recover higher initial costs on resale of building, but new energy rating systems are being introduced gradually.
- While the individual house may use an average of net zero energy over a year, it may demand energy at the time when peak demand for the grid occurs. In such a case, the capacity of the grid must still provide electricity to all loads. Therefore, a ZEB may not reduce the required power plant capacity.
- Without an optimised thermal envelope the embodied energy, heating and cooling energy and resource usage is higher than needed. ZEB by definition do not mandate a minimum heating and cooling performance level thus allowing oversized renewable energy systems to fill the energy gap.
- Solar energy capture using the house envelope only works in locations unobstructed from the sun. The solar energy capture cannot be optimized in north (for northern hemisphere, or south for southern Hemisphere) facing shade, or wooded surroundings.

Zero energy building versus green building

- The goal of green building and sustainable architecture is to use resources more efficiently and reduce a building's negative impact on the environment.
- Zero energy buildings achieve one key green-building goal of completely or very significantly reducing energy use and greenhouse gas emissions for the life of the building.
- Zero energy buildings may or may not be considered "green" in all areas, such as reducing waste, using recycled building materials etc.
- However, zero energy, or net-zero buildings do tend to have a much lower ecological impact over the life of the building compared with other "green" buildings that require imported energy and/or fossil fuel to be habitable and meet the needs of occupants.

5.1.2 Green building

- Green building (also known as green construction or sustainable building) refers to both a structure and the application of processes that are environmentally responsible and resource-efficient throughout a building's life-cycle: from planning to design, construction, operation, maintenance, renovation, and demolition.
- This requires close cooperation of the contractor, the architects, the engineers, and the client at all project stages. The Green Building practice expands and complements the classical building design concerns of economy, utility, durability, and comfort.
- Leadership in Energy and Environmental Design (LEED) is a set of rating systems for the design, construction, operation, and maintenance of green buildings which was developed by the U.S. Green Building Council.

- Another certificate system that confirms the sustainability of buildings is the British BREEAM (Building Research Establishment Environmental Assessment Method) for buildings and large-scale developments.
- Green building is the practice of creating structures and using processes that are environmentally responsible and resource-efficient throughout a building's life-cycle from siting to design, construction, operation, maintenance, renovation and deconstruction.

Eco-friendly Construction: 8 Advantages of Green Building

- Low Maintenance and Operation Cost. Green buildings incorporate unique construction features that ensure efficient use of resources such water and energy.
- Energy Efficiency.
- Enhances Indoor Environment Quality.
- Water Efficiency.
- Better Health.
- Material Efficiency.
- Better Environment.
- Reduces Strain on Local Resources.
- Green Buildings provide financial benefits that conventional buildings do not. These benefits include energy and water savings, reduced waste, improved indoor environmental quality, greater employee comfort/productivity, reduced employee health costs and lower operations and maintenance costs..
- There nine "elements" are literally your green building blocks.
  - Site planning and design.
  - Affordable housing works best when residents have easy access to key services and transit.
  - Community.
  - Indoor Air Quality.
  - Energy.
  - Materials.
  - Waste.
  - Water.
  - Commissioning.
- A green building rating system is a tool that evaluates the performance of a building and its impact on the environment. It comprises a predefined set of criteria relating to the design, construction, and operations of green buildings.
- Some of the more common green construction practices include: Using sustainable building materials like recycled glass and steel, as well as renewable materials like bamboo and rubber. Installing energy-efficient windows and doors. Using lower-VOC (volatile organic compounds) paints and stains.
- Green buildings preserve precious natural resources and improve our quality of life. There are a number of features which can make a building 'green'. These include: Efficient use of energy, water and other resources.
- The objective of green building concept is to develop buildings which use the natural resources to the minimal at the time of construction as well as operation. Green buildings emphasize on the resource usage efficiency and also press upon the three R's – Reduce, Reuse and Recycle.
- A high-performance building is a building whose energy efficiency and environmental performance is substantially better than standard practice.7 Although green buildings, on average, use less energy than conventional buildings, energy efficiency remains elusive.

- Building materials typically considered to be 'green' include lumber from forests that have been certified to a third-party forest standard, rapidly renewable plant materials like bamboo and straw, dimension stone, recycled stone, recycled metal.
- Indian Green Building Council (IGBC) Green Factory Buildings is the first rating programme developed in India, exclusively for the industrial sector. It is based on accepted energy and environmental principles and strikes a balance between known established practices and emerging concepts.
- The LEED rating systems are grouped into five main categories: Building Design and Construction. Interior Design and Construction. Operations and Maintenance.
- Green Architecture is largely driven by four goals:
- Reducing impacts on the Earth from constructing buildings and their materials (mining and harvesting, embodied energy, pollutants in manufacturing).
- Reducing impacts which arise during occupancy (fuel use, land pattern disruption, maintenance impacts).
- Eco-friendly, or ecological, construction is building a structure that is beneficial or non-harmful to the environment, and resource efficient. Eco-friendly construction has developed in response to the knowledge that buildings have an often negative impact upon our environment and our natural resources.
- Earth plaster and milk paints are excellent eco-friendly choices for building materials. Earth plaster is made out of mud and can be used as an interior or exterior finish. ... There are some great choices for eco-friendly flooring; cork flooring, bamboo, and recycled wood flooring are all excellent options.

  Advantages of green technology to the environment are enormous, but the most important benefits are:

  - It uses renewable resources that never depletes in nature.
  - Environmental technology uses latest energy generation techniques.
  - Offers environmental solutions for disposal and recycling of waste.
- Some examples of renewable energy sources are solar energy, wind energy, hydropower, geothermal energy, and biomass energy. These types of energy sources are different from fossil fuels, such as coal, oil, and natural gas.

5.1.3 Mass Housing

- Mass Housing is an independent public authority that provides financing for the construction and preservation of affordable rental housing, and for affordable first and second mortgages for homebuyers and homeowners.
- With Mass Housing's new refinancing product, home owners can refinance up to 100% of the appraised value of their home, pay-off existing approved second mortgages, have the option of paying no points, and roll-in closing costs.
- With Mass Housing's new refinancing product, home owners can refinance up to 100% of the appraised value of their home, pay-off existing approved second mortgages, have the option of paying no points, and roll-in closing costs.
- Today mass housing has become synonymous with low cost housing and the entire national effort in this field has been diverted to producing a cheaper house.
- Yet the numbers involved are of such magnitude that any amount of cost reduction exercises cannot provide the solution for housing everybody.
- Traditionally the individual who has built his own house has been intricately involved in the whole process. This made it possible for his identity to be established in his environment.
- Today the pace of development has taken away this close tie between the individual and the creation of his own environment. Large numbers of people are "designed for" by centralised agencies.

- This can cause further alienation unless the centralised agency concentrates on understanding the new demands of the housing activity and makes the process responsive once again to the individual needs. The task is, therefore, to evolve design strategies which allow the individual to participate once again in the building process.
- The fundamental requirements of shelter are the same for rich and poor. Everyone needs light, ventilation, healthy and clean surroundings, access to facilities etc.
- Whatever be the solutions to reduce cost, we have to find means to strengthen individual identity within the collective, and to provide the minimum environmental necessities.
- The enclosed programme provides a basis by which a pilot study can be carried out in Delhi. It shows how the existing housing process can be revolutionised by introducing "People's Participation" in mass housing.

5.1.3.1 Precast housing

- The concept of precast (also known as "prefabricated") construction includes those buildings where the majority of structural components are standardized and produced in plants in a location away from the building, and then transported to the site for assembly.
- These components are manufactured by industrial methods based on mass production in order to build a large number of buildings in a short time at low cost.

The main features of this construction process are as follows:

- The division and specialization of the human workforce
- The use of tools, machinery, and other equipment, usually automated, in the production of standard, interchangeable parts and products
- This type of construction requires a restructuring of the entire conventional construction process to enable interaction between the design phase and production planning in order to improve and speed up the construction. One of the key premises for achieving that objective is to design buildings with a regular configuration in plan and elevation.
- Prefabricated construction is the practice of assembling a variety of components of a structure at a manufacturing site and transporting those sub-assemblies to the location of the construction jobsite. Prefabricated construction is sometimes thought of as a low-end and mass produced mode of construction.
- Costs of prefabs are typically cheaper than stick-built homes by an average of 10% to 20%. Faster Than Traditional Construction. Generally, it takes one to four months longer to build a traditional home than to construct a prefab. Flexible Size and Aesthetic.
- A modular home is not a mobile home; it is simply a home that is built off-site, as opposed to on-site. These homes are often called factory-built, system-built or prefab (short for prefabricated) homes. Modular and manufactured homes are not the same. Manufactured homes are not placed on permanent foundations.
- Precast is a construction product (mostly concrete) produced by casting concrete in a reusable mold or "form" which is then cured in a controlled environment, transported to the construction site and lifted into place. In contrast, standard concrete is poured into site-specific forms and cured on site.
- The study revealed that, on average, the precast concrete slabs were 23.22% cheaper than cast-in-place concrete construction and precast columns were 21.4% cheaper than cast-in-place concrete construction, especially for large scale production.
- Precast concrete is known for its high-quality and durability. .One of the precast concrete advantages lies in its fireproof attributes. Precast concrete will not catch fire, and it will prevent the spreading of fire from room to room. It is a fantastic material to have a home or building made from for this very reason.

- Precast or Prefabricated construction concept includes those buildings where the majority of structural components are standardized and produced in plants or yards at a location nearer to the construction site; transported to building area for assembly or erection. These precast building components are manufactured in order to cater large number of buildings in a short duration at lower cost.
- The main features of precast construction process are as follows: The use of standard tools, machinery, and equipment in the production of standard, interchangeable parts and products.
- In urban areas, residential buildings of this type are usually five to ten stories high or more.
- Many countries used various precast building systems to provide low-income housing for the growing urban population. In general, precast building systems are proved to speed-up the project with better quality control; which leads to economy when compared to conventional method of construction (apartment buildings) in many countries.
- Precast construction method is purely based on shear wall concept and hence the typical floor doesn't contain any type of columns but in conventional, columns are must to construct twelve floors of a tower.
- Precast construction process carried out at site as production of elements along with simultaneous erection of the ready elements which considerably saves the time of construction where as in conventional, block work and curing consume much time to complete.
- Block work is limited to only 1 to 1.2 m height per day to maintain proper alignment whereas; precast elements do not have any such restrictions.
- In conventional method, block work must proceed further with proper curing and also plastering require curing period for sufficient hydration and to attain strength; but in precast method, no such intense curing required because, the main and final curing done at casting yard only to achieve maximum strength.
- Plastering work is mandatory in conventional whereas precast method does not require any kind of plastering work.
- For slabs, shuttering work, reinforcement works consume more time in conventional method; but in precast, slabs also get manufactured in yard itself.
- All parts of a structure cannot be fixed with precast elements; at few necessary locations block work is essential in precast construction also.
- In precast method, slab should proceed further with screed of certain thickness to provide necessary electrical conduits; whereas conventional system contains the electrical conduit works in normal slab shuttering work itself.
- No alterations in design are possible in precast construction; but in conventional, any kind of alterations are made without any problem.
- Better quality control can be possible with precast construction but as the floor level increases, in conventional system it is very difficult to maintain good quality.
- Precast construction system requires almost half the number of total man power that required for conventional system.
- Construction activities take place at a faster rate in precast compared to conventional system.
- Precast construction provides better alignment of structural elements than the conventional one.
- Precast construction is best suitable for large construction projects; whereas conventional system lead to various problems and consumes huge time for completion.

5.1.3.2 Prefab Homes

- Prefabricated homes, often referred to as prefab homes or simply prefabs, are specialist dwelling types of prefabricated building, which are manufactured off-site in advance, usually in standard sections that can be easily shipped and assembled.

- "Prefabricated" may refer to buildings built in components (for example, panels), modules (modular homes) or transportable sections (manufactured homes), and may also be used to refer to mobile homes, i.e., houses on wheels.
- Although similar, the methods and design of the three vary widely. There are two-level home plans, as well as custom home plans. There are considerable differences in the construction types.
- Modular homes are created in sections, and then transported to the home site for construction and installation. These are typically installed and treated like a regular house, for financing, appraisal and construction purposes, and are usually the most expensive of the three.
- Although the sections of the house are prefabricated, the sections, or modules, are put together at the construction much like a typical home.
- Manufactured homes, once placed on a permanent foundation, are considered the same as modular or site build homes for appraisal purposes.
- Manufactured homes are built onto steel beams, and are transported in complete sections to the home site, where they are assembled. Wheels, hitch and axles are removed on site when the home is placed on a permanent foundation.
- Mobile homes, or trailers, are built on wheels, and can be pulled by a vehicle. They are considered to be personal property, and are licensed by the Dept. of Motor Vehicles.
- Tiny homes with wheels are included in this category. They must be built to the DMV code, and pass inspection for licensing.
- Constructing manufactured homes typically involves connecting plumbing and electrical lines across the sections, and sealing the sections together. Manufactured homes can be single, double or triple-wide, describing how many sections wide it is.
- Many manufactured home companies manufacture a variety of different designs, and many of the floorplans are available online. Manufactured homes can be built onto a permanent foundation, and if designed correctly, can be difficult to distinguish from a stick-built home to the untrained eye.
- A manufactured home, once assembled, goes through a 'settling-in' period, where the home will settle into its location. During this period, some drywall cracking may appear, and any incorrectly installed appliances, wiring or plumbing should be repaired, hopefully under warranty.
- If not covered under warranty, the costs will be borne by the consumer. For this reason, it is important that the consumer ensure that a reputable and honest contractor is used for the initial set-up.
- If any repairs are not completed by the initial set-up crew, the manufacturer will send repair crews to repair anything covered by the warranty.
- The secondary repair team must be scheduled, and may not be available immediately for most repairs. Just because a manufactured home has been assembled does not mean it is immediately habitable; appropriate ventilation, heating, plumbing, and electrical systems must be installed by a set-up crew, otherwise, the buyer must wait for the manufacturer repair team or do it themselves.

Advantages of Prefab Homes

- **Modernizing the Look with Energy Efficiency:** Though modular homes are historically associated with lower quality, this is no longer the case today. In order to stand out from the competitive modular market, builders are looking to maximize energy efficiency through numerous methods such as recycling materials, using LED lighting and installing solar panels. This ensures that you save money on your energy bills to the point where you may be producing clean energy that will be put back into the energy grid.

- **Zero Waste and Zero Inspection Hassle:** After being created in the factory setting, these modular pieces are then transported to site and assembled using heavy machinery. Similar to a BONE Structure home, modular homes do not produce waste as all the required materials for construction arrive on site already installed in the modular pieces. It would be like putting together a Lego home – except all the pieces are rooms, designed to your desire and needs. In addition to their ease of construction and energy advantages, modular homes are almost always manufactured to code, ridding you of the worry that comes with home inspections by your municipality.
- **Reliable Schedule:** An additional advantage to modular homes is the minimal opportunity for negative schedule interference to occur. This is in part due to the construction of the pieces being in a controlled indoor environment that can't blame weather for delays. Without weather interruptions, the average home would be constructed in no longer than four months. During this construction process, workers are put in a low risk environment that elevates their feelings of comfort. With healthier workers, you can be certain that your home will receive the care and attention to details that it requires in order to be built safely and soundly.

Disadvantages of Prefab Homes

- **Limited Design Options:** On the flip side – modular homes are known to come with their own sets of disadvantages. Depending on the company you choose to build your modular home with – you can be stuck with a limited amount of material options and home layout possibilities. This implies that your home may have less flexibility in the design than you anticipated.
- **Reduced Resell Value:** Modular homes also have a stigma surrounding them: that they are of lower quality, which makes them extremely difficult to resell. Modular buildings have historically been associated with lower quality homes that boast dated designs, such as every home buyer's nightmare: popcorn ceilings.
- **Difficult to Finance:** Lastly, modular homes need to be built with a corresponding finance plan that will differ from the mortgage plan associated with traditional stick frame homes. Banks are generally unfamiliar with the modular home construction process and the fact that most payments are required to be made upfront. Banks have been known to deny some people the mortgage required to support this process and clients have had to look at various options before being able to continue with the construction of their modular home.

5.1.3.3 Pre-engineering building

- Pre Engineered Buildings (PEB) is the buildings which are engineered at a factory and assembled at site. Usually PEBs are steel structures. Built-up sections are fabricated at the factory to exact size, transported to site and assembled at site with bolted connections. This type of Structural Concept is generally used to build Industrial Buildings, Metro Stations, and Warehouses etc.
- Pre-Engineered Metal Building Construction is a cost-efficient alternative to conventional building design in commercial applications is the pre-engineered metal building system. Metal buildings consist of a structural steel framing system, supporting a metal roofing system and the wall panels of varying materials.
- Pre-Engineered Metal Building Construction. A cost-efficient alternative to conventional building design in commercial applications is the pre-engineered metal building system. Metal buildings consist of a structural steel framing system, supporting a metal roofing system and the wall panels of varying materials.
- Pre-engineered buildings are the new contemporary structures ideal for commercial purposes. PEB structures can accommodate industrial units that are used for manufacturing of consumer goods and products. Pre-engineered steel buildings are constructed at low costs and can be used for industrial applications extensively.

- The adoptability of PEB in the place of Conventional Steel Building design concept resulted in many advantages, including economy and easier fabrication. This type of building structure can be finished internally to serve any function that is actually help in low rise building design. Examples of Pre-Engineered Buildings are warehouses, Canopies, Factories, Bridges etc.

Components of PEB:

Pre Engineered Buildings are consist of different steel structural member which are as follows,

- Primary Frame: Primary framing of a PEB is an assembly of builtup I-Shaped steel members and that framing consist trusses or castellated beams etc.
- Secondary Structural Elements: It is actually Cold Formed Members, which can be in diff. shapes like "Z", "C" etc. In general known as "Purlins".
- Roof and Wall Panels : Tin shades and Curtain Wall made of Glass and Roll-formed steel sheets usually comes in this category , S.
- Sandwich Panels: Sandwich Panel is made of three layers, in which a non-Aluminum Core is inserted b/w two aluminum sheet.
- Other Accessories: Mezzanine floors, Bolts, Insulation etc.

Advantages of PEB:

- There are many advantages of Pre Engineered Buildings, which are as follows,
- Quality control is the main advantage as all the structural member are engineered beforehand, standards of different codes also taken into consideration and these components are made in factory under the supervision of Quality Control Engineer.
- Lower cost due to the saving in design, manufacturing and on site erection cost.
- Minimizing time of construction due to the use of software for design of the structural components.
- Low Maintenance due to use of standard quality of paints over steel members , which increases the ability to withstand and finally the maintenance cost will be low as compare to conventional steel building.
- Quick Erection, as all the members are Pre Manufactured and skilled labor is used for connections of different components.
- Warranty on PEB, mostly warranty period of 20 years given by manufactures for PEB.

Disadvantages of PEB:

- PEB have many advantages in the field of Industrial structure but still there are some demerits of Pre Engineered Buildings, which are as follows,
- Rusting/Corrosion Sensitive, as if the quality of steel used or paint used for coating of steel members is not of good quality, than it can damage the structure and thus reduces the life of structure.
- Insulation Cost, as insulating the building to an agreeable benchmark will furthermore add to your construction costs.
- Appearance, Steel Sections can be unattractive when left exposed.

5.1.4 Solar Paints

- Solar paint, also known as photovoltaic paint, is exactly what it sounds like! It's a paint that you can apply to any surface that will capture energy from the sun and transform it into electricity. The paint would essentially be your average paint, but with billions of pieces of light sensitive material suspended in it, material that would transform the typical paint into superpowered energy-capturing paint.
- This new photovoltaic technology is Solar paint. The new paint is composed of synthetic molybdenum sulphide.

- The paint will be based on dye-sensitized solar cells. Instead of absorbing sunlight using silicon like conventional solar panels, they use dye molecules attached to particles of the titanium dioxide pigment used in paints. That gives an energy boost to electrons, which hop from the dye into a layer of electrolyte.
- Solar paint, also known as photovoltaic paint, is exactly what it sounds like! It's a paint that you can apply to any surface that will capture energy from the sun and transform it into electricity.
- Right now, solar paint is a little bit more than a concept and a little bit less than a consumer-level product.
- The main issue with solar paint is that it's not quite efficient enough to be commercially viable.
- Solar paint ranges anywhere from 3-8% Efficiency is essentially the percentage of the power of the sun's rays that the technology can capture.
- If a particular type of solar paint has a 5% efficiency that means it's capturing only 5% of the total available sun energy. For comparison sake, traditional silicon solar panels operate at around 18% efficiency.
- Most experts agree that a solar technology has to surpass 10% efficiency to be viable.
- Solar paint is certainly less expensive than solar panels, in terms of both production and installation, but the low efficiency means that it's not quite ready to be taken to market.
- However, we still think it's quite amazing that the technology exists at all. The widespread usage of solar paint is not as far off as it may seem!
- The most common type of photovoltaic paint is a paint utilizing *colloidal quantum dots.* These are semiconductor crystals that are already used in solar panels as well as LEDs and computers.
- This backing could then be rolled-up, sent to the place where it's to be installed, and then applied like wallpaper. The only issue with the dots technology is that the efficiency is still sitting around 8% and therefore not yet commercially viable.
- The solar industry is a progressive industry that is always looking to improve its technology.

5.1.5 Building Integrated Photovoltaic (BIPV)

- Acronym of BIPV (Building Integrated Photovoltaics) refers to photovoltaic systems integrated within an object. In the case of facade or roof systems the photovoltaic system is added to the building after it was built. These low powered systems of up to some 10 kW are usually integrated into the south facade.
- Photovoltaic glass (PV glass) is a technology that enables the conversion of light into electricity. To do so, the glass incorporates transparent semiconductor-based photovoltaic cells, which are also known as solar cells. The cells are sandwiched between two sheets of glass.
- Solar panels are made of tempered glass, which is also referred to as safety glass or toughened glass. It is created by thermal or chemical means. There are specific characteristics that make tempered glass suitable for the manufacturing of solar PV panels.
- Building-integrated photovoltaics (BIPV) are photovoltaic materials that are used to replace conventional building materials in parts of the building envelope such as the roof, skylights, or facades.
- They are increasingly being incorporated into the construction of new buildings as a principal or ancillary source of electrical power, although existing buildings may be retrofitted with similar technology.
- The advantage of integrated photovoltaics over more common non-integrated systems is that the initial cost can be offset by reducing the amount spent on building materials and labor that would normally be used to construct the part of the building that the BIPV modules replace. These advantages make BIPV one of the fastest growing segments of the photovoltaic industry.

- There are four main types of BIPV products:
  - Crystalline silicon solar panels for ground-based and rooftop power plant.
  - Amorphous crystalline silicon thin film solar pv modules which could be hollow, light, red blue yellow, as glass curtain wall and transparent skylight.
  - CIGS-based (Copper Indium Gallium Selenide) thin film cells on flexible modules laminated to the building envelope element or the CIGS cells are mounted directly onto the building envelope substrate.
  - Double glass solar panels with square cells inside.
- Building-integrated photovoltaic (BIPV) electric power systems not only produce electricity, they are also part of the building. For example, a BIPV skylight is an integral component of the building envelope as well as a solar electric energy system that generates electricity for the building. These solar systems are thus multifunctional construction materials.
- The standard element of a BIPV system is the PV module. Individual solar cells are interconnected and encapsulated on various materials to form a module. Modules are strung together in an electrical series with cables and wires to form a PV array. Direct or diffuse light (usually sunlight) shining on the solar cells induces the photovoltaic effect, generating unregulated DC electric power.
- This DC power can be used, stored in a battery system, or fed into an inverter that transforms and synchronizes the power into AC electricity. The electricity can be used in the building or exported to a utility company through a grid interconnection.
- A wide variety of BIPV systems are available in today's markets. Most of them can be grouped into two main categories: facade systems and roofing systems. Facade systems include curtain wall products, spandrel panels, and glazings.
- Roofing systems include tiles, shingles, standing seam products, and skylights. This sourcebook illustrates how PV modules can be designed as aesthetically integrated building components (such as awnings) and as entire structures (such as bus shelters).
- BIPV is sometimes the optimal method of installing renewable energy systems in urban, built-up areas where undeveloped land is both scarce and expensive. The fundamental first step in any BIPV application is to maximize energy efficiency within the building's energy demand or load.
- This way, the entire energy system can be optimized. Holistically designed BIPV systems will reduce a building's energy demand from the electric utility grid while generating electricity on site and performing as the weathering skin of the building.
- Roof and wall systems can provide R-value to diminish undesired thermal transference. Windows, skylights, and facade shelves can be designed to increase day lighting opportunities in interior spaces. PV awnings can be designed to reduce unwanted glare and heat gain. This integrated approach, which brings together energy conservation, energy efficiency, building envelope design, and PV technology and placement, maximizes energy savings and makes the most of opportunities to use BIPV systems.

Application

Photovoltaics may be integrated into many different assemblies within a building envelope:

- Solar cells can be incorporated into the facade of a building, complementing or replacing traditional view or spandrel glass. Often, these installations are vertical, reducing access to available solar resources, but the large surface area of buildings can help compensate for the reduced power.
- Photovoltaics may be incorporated into awnings and saw-tooth designs on a building facade. These increase access to direct sunlight while providing additional architectural benefits such as passive shading.
- The use of PV in roofing systems can provide a direct replacement for batten and seam metal roofing and traditional 3-tab asphalt shingles.
- Using PV for skylight systems can be both an economical use of PV and an exciting design feature.

5.1.6 Earthquake Resisting Controls

- Seismic isolation and energy dissipating systems present an effective way to common seismic design for improving the seismic performance of structures. These techniques reduce the seismic forces by changing the stiffness and/or damping in the structures, whereas conventional seismic design is required for an additional strength and ductility to resist seismic forces.
- Earthquake engineering include the analysis and evaluation of earthquake risk and the analysis and design of different types of engineering structures.
- Strength and ductility that are key consideration in earthquake resistant design.
- The purpose of earthquake prevention of buildings is to provide the structural safety and comfort by controlling the internal forces and displacement within the particular limits.
- The common method for protecting the structures against the destructive effects of earthquakes is to damp the seismic energy for limiting the seismic energy by the structural elements, thus providing the resistance against the earthquake.
- In spite of using this method for a certain level of protection, the structure could be damaged for real sometimes.
- Another method for protection of the structures against the earthquake is to **isolate** the building from the ground and/or to install seismic energy **dissipating** elements at the appropriate places of the building.
- With this method, better protection could be provided, by designing correctly against the earthquake and therefore significant structural damage level could be minimized.
- The earthquakes have been carried on to be an important factor that threatens the social and economic future of the countries, as we can observe the results of them. Thus, it is insisted on the resolutions that minimize the seismic effects of the buildings should demonstrate a high performance level in the expected earthquakes.
- The seismic isolators and energy dissipating devices are seen to be effective solutions within this context, which are placed in the building appropriately to damp the seismic energy or placed between the foundation and vertical structural systems damping the seismic energy under the ground of the building, thus decreasing the effects of lateral loads on top floors.
- Application of earthquake protection systems in buildings whether will be constructed and were constructed-especially the historical ones-, increases the importance of these technologies.

5.1.6.1 Isolation

- All the operations that are developed in order to protect the buildings against the earthquakes, thus providing the security and comfort conditions under service loads and that include placing certain kind of additional elements in the building are named seismic isolation in general. Various kinds of devices are used in the buildings for the purpose of seismic isolation.
- It would be useful to examine the seismic isolation in the framework of the basic principles of dynamics before introducing these devices. Seismic isolation in a building, when considered within the framework of basic principles of dynamics, can be maintained by taking under the control, modifying and changing the characteristics of both restoring-force when affected by seismic forces, and damping of the building, and also the mass of the building and seismic forces that affect the building.
- As it is known, the equation of motion of a building that is subjected to the ground motion depends on mass, stiffness, and energy damping nature of the building, as well as on external seismic forces affecting the building.
- The characteristics of response forces can be controlled, by changing stiffness of the building. When stiffness of the building is decreased, the response acceleration also decreases and displacements increase.

- On the other hand, response of acceleration and displacement can be decreased, by increasing the damping effect of the building. Various kinds of dampers and their combinations can be placed in the building. The dynamic characteristics of the structural system might be varied, by changing total mass and the distribution of the mass within the system.
- Controlling and arranging the seismic forces that affect the building can be achieved by isolating the building from the ground.
- A "smart" base isolation strategy is proposed and shown to effectively protect structures against extreme earthquakes without sacrificing performance during the more frequent, moderate seismic events. The proposed smart base isolation system is composed of conventional low-damping elastomeric bearings and "smart" controllable (semi active) dampers, such as magneto rheological fluid dampers. To demonstrate the advantages of this approach, the smart isolation system is compared to lead-rubber bearing isolation systems. The effectiveness of the isolation approaches is judged based on computed responses to several historical earthquakes scaled to various magnitudes.

Clasification of seismic isolation systems

1. **Active Control Systems**
   - The operation mechanism of active control systems is based on providing a continuous energy from outside. That is why the cost of setting up these systems is high. The system can control the acceleration, displacement or velocity of the structure.
   - Active control systems are composed of electronic devices such as computers, actuators and starters. The design of active control systems is independent from the intensity of the ground motion in the following way.
   - The system changes its rigidity or the quantity of motion according to the intensity of the ground motion. Therefore, the factors such as the uncertainty or unpredictability of the future ground motions are not important in the designs that make use of active control systems.
2. **Passive Control Systems**
   - Passive control systems operate without utilization of any external energy source. Therefore, the cost of these systems' setting up is less in comparison with active systems.
   - These systems can control the displacement up to a certain limit. In the passive control systems, the protection systems are designed in accordance to protection level required for earthquakes of certain magnitude.
   - These systems are composed of dampers, isolators and other devices that can easily be found and applied.

5.1.6.2 Dissipation

- Another approach for controlling seismic damage in buildings and improving their seismic performance is by installing Seismic Dampers in place of structural elements, such as diagonal braces. These dampers act like the hydraulic shock absorbers in cars - much of the sudden jerks are absorbed in the hydraulic fluids and only little is transmitted above to the chassis of the car. When seismic energy is transmitted through them, dampers absorb part of it, and thus damp the motion of the building.

Commonly used Seismic Dampers (Dissipation Devices)

- **Viscous Dampers** (energy is absorbed by silicone-based fluid passing between piston cylinder arrangement).
- **Friction Dampers** (energy is absorbed by surfaces with friction between them rubbing against each other).
- **Yielding Dampers** (energy is absorbed by metallic components that yield).
- **Viscoelastic Dampers** (energy is absorbed by utilizing the controlled shearing of solids).

New Breed of Energy Dissipation Devices

- The innovative methods for control of seismic vibrations such as frictional and other types of damping devices are important integral part of seismic isolation systems as they severe as a barrier against the penetration of seismic energy into the structure. In this concept, the dampers suppress the response of the isolated building relative to its base.
- The novel friction damper device consists of three steel plates rotating against each other in opposite directions. The steel plates are separated by two shims of friction pad material producing friction with steel plates.

5.2 ROAD CONSTRUCTION TECHNIQUES

- It is undeniable that the road construction industry every day is looking for a greater effectiveness and efficiency in its techniques and methods.
- However, change in the construction environment is not willingly embraced by the construction industry and this conflict to change compromises innovation performance, and unenthusiastically impacts customer and industry goals.
- In other words, with rising globalisation, roads have become a very vital infrastructure in enabling the transfer of freight as well as people, making the better and sustainable development of roads very important. Innovation in road construction offers important industry and community benefits.

5.2.1 3D Printing

- The idea of 3D printing of structures, although still novel to many, is striding its way towards the root of the construction industry.
- Today, 3D printers are being developed to process all kind of construction materials and are being tested and tinkered with every day for refinement.
- Just the thought of an immense reduction in the project span and cost with the use of 3D printing technology has been keeping everyone on their toes.
- 3D printing, also known as additive manufacturing or contour crafting, holds great potential in the industry when compared with the conventional methods of construction.
- The printing process, to explain in the simplest terms, involves the feeding of relevant data like maps, plans, blueprints and dimensions in an orderly and predefined way to an automated machine which in turn produces a precise physical output.
- 3D printing uses the SLA or stereolithography process which means the creation of a product is done in a layer-by-layer fashion.
- Building elements like walls, beams and vertical support structures can be 3D printed right at the site whereas the other elements like doors and windows, electrical and plumbing systems are to be installed with the help of conventional methods.

Process for 3D Printing in Construction

- A blueprint is prepared with the help of CAD software with all the necessary details in it. Since this is going to be the final design and the output would be based on this, precision of details is pivotal.
- The CAD file prepared is then converted into the ".stl" (standard tesselation language) extension which is compatible with the printer. The extension was first used in stereolithography in the year 1987.
- The orientation of the design and the printer is adjusted just as we do when using a 2D printer. Once set up, the cartridges are filled-up with the requisite powders and binders for the production of the structure.
- The printer, when given the command, would start creating the product, gradually building up the structure one microscopic layer at a time. The thickness of the paste extruded by the nozzle can be varied depending on the structure being printed. A typical layer is around 0.1 mm in thickness.

- While the machine is laying the paste, no intrusion is required; the printer will simply keep on layering the levels until the job has reached fruition. Periodic checks are all that are required to ensure everything is alright. The time period required for the completion of a job can range from several hours to days, depending on the size and complexity of the structure.

Benefits of 3D Printing

- **Quicker construction:** While construction through conventional methods can take months and years to complete a piece of work, 3D printing can wrap-up the same piece of work within days.
- **Improved efficiency:** Since 3D printing is an additive method, the material is added to create the final product contrary to the manufacturing processes such as CNC machining which are subtractive and leave out a lot of scrap and wastage.

Design flexibility

- 3D printing provides a great scope for innovation and creativity for the users. Intricate designs and curves which otherwise unimaginable can be built using this technique.

5.2.2 Road Printer

- Brick roads have been around us for centuries. They are both durable and treat for the eyes and easy to produce and reuse, cement pavers last a very long time and they are easy to repair and replace But Now-a-days there are difficult to find because building them is an arduous and back breaking task.
- Engineers are always busy in finding solutions to problem around the world and strive to decrease the Time consumption to complete a task. One such an endeavor has resulted in one such machinery which is more than just a symbol of great engineering it is both cool and awesome "THE TIGER STONE".
- As Belgian manufacture RPS, the machine caries ergonomic benefits for its operators: "because the finishing of the edges is directly done on the machine, the pavers do not have to get on their knees for fixing the edges "they work in a good ergonomic position which makes their job much lighter and more efficient and road is filled and finished instantly".

Machine:

- The tiger stone is a six meters wide machine that can repave entire street at once, including edges. The device is amazingly simple to handle. Any person can work with it within 5 minutes.
- Naturally the stones fall on the road directly into the appropriate pattern. The tiger stone is a solution for the heavy work that street pavers in Holland usually have to do manually.
- This machine is known as TIGER STONE paving machine and is being called 'road printer' due to the way it works.

Procedure

- The RPS6 ROAD PRINTER is a divisible paving machine capable of laying bricks for roads. As the machine roll backwards 'GRAVITY' brings the bricks flat to rest on the ground.
- A front loader fills the brick hopper with loose bricks, the bricklayers stand on the machine and assemble the pattern vertically.
- Tiger stone requires 1-3 operators on its platform to provide the machines pusher slot with looser bricks from the hopper.
- The green machine runs entirely on electricity, moving at a speed of 13 feet per a minute.
- Paving stones are lowered smoothly via a curved ramp into place in the correct pattern, it selfsteers to keep the road straight and in place.
- Then tamper (or) steamroller pushes the bricks down to secure them in place.
- Bricks roads are experiencing a revival, because they can be reused easily, filter water and without expanding ice.

- The road printer is capable of creating roads anywhere from 3-20 feet wide.
- As construction workers insert brick into the machine's forward compartment, road printer automatically processes the bricks and splits the assembled road from its rear.

Advantage

- The tiger stone brick printer makes road constructions as easy as laying laminate flooring.
- Tiger stone is powered electrically and has very little moving parts therefore the noise and maintenance is quite less.
- It can lay down about 400 yards of road per day.

Disadvantage

- The cost of the equipment is expensive.

5.2.3 Smart roads

- Smart road are terms for a number of different ways technologies are incorporated into roads, for improving the operation of connected and autonomous vehicles for traffic lights and street lighting, and for monitoring the condition of the road, traffic levels and the speed of vehicles.
- Intelligent transportation systems usually refers to the use of information and communication technologies (rather than innovations in the construction of the roadway) in the field of road transport, including infrastructure, vehicles and users, and in traffic management and mobility management, as well as for interfaces with other modes of transport.
- Vehicle and highway automation is believed to reduce the risk of accidents, improve safety, increase capacity, reduce fuel consumption and enhance overall comfort and performance for drivers. There has been enough reason to assume that more automated automobiles relieve the driver from many undesirable routines of driving task.
- It has also been known that many of the car accidents are due to human errors. Therefore, the conclusion has been that with a robust automated system the chance of car accidents can be reduced. With the overwhelming increase in the number of vehicles on the road another concern has been road capacity.
- Some kind of automation that would help to safely increase traffic flow has been considered as one potential solution to Congested Highways a smoother cruise with an automated system can reduce fuel consumption and engine wear.
- The potential of connected, smart roads is huge. Not only will they keep us safe by regulating the speed of our vehicles and implementing warning systems, but also transmit real time data and share information across the network, making it simpler and quicker to get around, to find parking, to commute effectively and communicate with each other.
- The principal idea of solar road panels is to utilise the space occupied by roads to generate electricity via photo-voltaic panels installed in place of a conventional concrete or asphalt road surface.

Some Major idea include in smart roads:

1. Solar roads.
2. Self-healing concrete roads.
3. Electric charging stations.
4. Road police drones.
5. Weather and traffic detection.

Other functions for solar road panels have been proposed, including the following:

- The panels could include LED lights for creating dynamic road markings, for examlple, lane markings, or warning messages such as "Reduce Speed".
- The panels could include heating elements that produce sufficient energy to clear ice and snow from roadways.
- The panels could include wireless charging technology to recharge the batteries of electric vehicles that drive over the panels.

5.3 COASTAL CONSTRUCTION TECHNIQUES

- The coastal technologies in the inventory have been structured according to the components in a building.
  - Wall Technologies.
  - Roof Technologies.
  - Construction Systems.
  - Flooring.
  - Energy systems.
- The inventory includes both alternate materials and technologies that offer benefits in terms of the environment and economics, while maintaining efficiency and quality parameters.
- The technologies have been chosen based on the following key parameters.
  - Reduced embodied energy and fuel consumption hence reduced carbon emissions.
  - Reduced environmental damage through optimal resource use and waste utilization.
  - Better Thermal efficiency and comfort.
  - Resistant to natural disasters.
  - Aligned to local production in terms of material and skill availability.
  - Cost efficiency.

5.3.1 Sound Proofing walls

- For sound proof and fire proof inner walls, of a type comprising two parallel reinforcing metal plates mechanically interconnected to form an interspace there between and a filler of insulating material, constituted essentially of rock-wool like material arranged in the interspace. At least one continuous diaphragm extends parallel to the reinforcing metal plates and divides the filler into parallel layers. The diaphragm comprises a plate of solid refractory material having projecting edges projecting beyond the filler. The reinforcing metal plates have folded edges constituting section member like portions thereof. The projecting edges are in engagement with the folded edges of the metal plates. The folded edges are seam folded to hold the metal plates together.
- Sound proofing materials can be applied to a wall that's under construction or be added to an existing wall, post-construction. There are several ways to sound proof a wall. Improvements in the sound proofing industry have resulted in products that are more effective and cost less than traditional wall soundproofing methods.
- In almost every way, the modern drywall-over-studs wall is better than its timber-and-masonry and plaster-and-lath ancestors. It is fast and easy to build, lightweight and makes the most of inexpensive materials. But when it comes to stopping sound, the modern wall is a flop.
- The sound proofing walls process involves ripping the existing drywall-off the walls (and perhaps the ceiling), filling the walls with fiberglass insulation, attaching metal strips called "resilient channel" to the studs, and fastening new drywall to the channel.
- This straightforward how to make your room soundproof project doesn't require specialized tools or high-level construction skills. Anyone who has experience hanging and taping drywall, along with a little carpentry and electrical know-how, can do this job to sound proof a room.
- Soundproofing is a messy, labor-intensive project, however. To minimize household havoc, it's best to focus on one room, or at least one room at a time. The room might be a place that you want to keep sound out of – a home office, for example. Or it may be a room you want to keep sound in – like a home theater.
- We won't show you how to block noise coming from outside since most exterior walls already block sound fairly well. And any improvement you make to them will be of marginal benefit unless you also upgrade your windows.

Recommended Wall Soundproofing Methods

Recommended soundproofing methods are less expensive and more effective than traditional methods.

- Adding Mass.
- Recycled Cotton Batts.
- Sound Isolation Systems.

Traditional, Less Effective Methods

- Traditional wall soundproofing methods are more expensive and less effective than the new technologies available today.
- Staggered Stud Walls.
- Double Walls.
- Resilient Channel System.
- Fiberglass Batts.
- Expanded Foam Sprays.

5.3.2 Water Resistant Roofs

- Waterproofing requires you to block moisture out by using techniques to seal your roof against water.
- If you have a low slope roof or one with many architectural elements such as multiple gables and valleys, waterproofing may be a necessity.
- A roof with a low slope may not shed water efficiently and experience pooled or standing water for a period of time. Water will always seek to run downward; if there is an opening, it will get through and create water damage to anything underneath the roof, from the decking and insulation to the building's contents.
- If water gets trapped in the end lap or the purlin, it can cause premature corrosion in the secondary framing members or panel.
- A roof with a slope of less than 3:12 is generally considered to have a low slope.
- If there are panels or areas of the roof what are slightly lower than the surrounding materials, it will hold standing water.
- Other places water sneaks in is through exposed fasteners and through nesting panels, which may experience wicking, drawing water underneath the edge.

Waterproofing is best approached via. multiple aspects of the roof.

- The most water resistant roof is a standing seam roof with seams at least two inches high. Ensure the roof can move with thermal expansion and contraction without damage.
- Mechanical seaming of the joints between panels eliminates areas where water may enter between panels.
- Do not use exposed fasteners. Opt for concealed fasteners for the entire roof. Fasteners do not always last the life of the roof and become a primary entry point for moisture.
- Use foam and metal closures along with manufacturer approved sealant to ensure a watertight fit between panels, at the end laps and around fasteners. Do not allow the sealant to be exposed to the outside environment as it will shorten the life of the material.
- Include proper ventilation to protect against condensation on the underside of metal panels and trim.
- The waterproofing membrane is considered to be most important component of the roofing system as it serves the main function of keeping water out of the building. Below are some common roof waterproofing membranes used locally:

(a) Liquid-applied membranes

- Liquid-applied membranes are applied on site in a liquid form which is allowed to set and form into a water impermeable membrane. Bituminous-based (except those containing coal tar) and polymeric-based membranes can be applied when they are cold while those containing coal tar are usually applied when heated. Liquid-applied membranes are seamless, semi-flexible, easy to apply, detail, maintain and repair. However careful supervision and control during application is needed, particularly in ensuring proper curing of concrete, consistent thickness and uniform application.

(b) Pre-formed sheet membranes

- One type of pre-formed membranes is the polymer-modified bitumen membranes and they are applied by heat or attached with an adhesive. These blended or 'modified' asphaltic product are bonded to a high strength fabric of polyester or fiberglass and produced into rolls. They have elongation and recovery properties which make them suitable to protect against stresses created by wind, temperature fluctuation and normal structural expansion and contraction of the building.
- Another type of pre-formed membrane is made of Polyvinyl Chloride (PVC). Although PVC is a hard resin, it is modified with the addition of plasticisers to make it more supple and pliable for use as roofing membranes.

5.3.3 High Performance Doors and Windows

- High Performance Doors are automatic doors used for fast access between internal and external areas of buildings.
- Prime High Performance Doors are designed and factory-made with German collaboration using state-of-the-art and resourceful engineering technology for frequent use in high traffic areas with opening speeds up to 3 m/s.
- The high opening speed reduces time taken in movement of material, thus hastening the logistics process. Our team of skilled engineers visits site and recommend the most suitable rapid door to client. Rapid doors are manufactured and installed keeping in mind international standards.
- Many manufacturing industries nowadays need a controlled environment in which you limit the amount of dirt and dust in the manufacturing premises. Medical instrument manufacturer, electronics and computer manufacturer, food industry, pharmaceutical industry and some military applications are but a few of the examples that have strict requirements for maintaining dust free environment. High speed doors have become integral part of every industry.
- A wide range of Prime High Speed Doors are available for various operations, including areas with special requirements for temperature control, hygiene, storage and handling of frozen foods, Atex rapid doors are need in explosion proof areas, deep freezer automatic roll-up door are need in cold storages, warehousing, loading bays and on conveyor systems. Prime Reset features a special advanced self-repairing system.
- The present invention discloses a composite profile for thermal bridge solid-web steel doors and windows and a manufacturing method, relating to a steel door and window profile and a manufacturing method. The profile provided by the invention consists of a solid-web steel profile made from standard steel member and a nonmetallic heat insulation strip.
- The manufacturing method is that: firstly one to seven kinds of standard steel members are manufactured; secondly two standard members are seam welded into eight kinds of solid-web steel profiles; then two kinds of the solid-web steel profiles and one nonmetallic heat insulation strip are assembled into six kinds of composite profiles for thermal bridge solid-web steel doors and windows; and finally the coating and the assembly are carried out.

- Due to the small cross section, light weight, high strength, steel material saving and convenient assembly of the steel member, the profile provided by the present invention has the advantages of good heat insulation performance and high quality. The present invention also has the advantages of simple method, low equipment investment, standardized manufacture of the steel member, high efficiency and convenient popularization and application.
- Thus the present invention can be widely applied in the composite sections for thermal bridge solid-web steel doors and windows for the flat-opening door and window, the sliding door and window, the fixed window, the top-hinged swinging window and the manufacture.
- The most efficient HVAC or lighting system is the one that does not have to work at all because the envelope has neutralized or captured the thermal load at the building skin and provided required interior illumination. These high-performance energy functions can be implemented while also providing view, security, thermal comfort, privacy, and aesthetic solutions demanded of building owners, thereby enhancing the economic attractiveness of these solutions and accelerating widespread market adoption.

High Performance Windows

- Vacuum Insulated Glazing (VIG).
- Dynamic Windows.
- Thin-Glass Triple Center Layer.
- Center-of-Glass Deflection.
- Operating Hardware.
- Highly Insulating Commercial Framing.
- Gas Distillation.
- Energy Recovery Based Facade Ventilation Systems.

5.3.4 Air and Moisture Barriers

- The job of a moisture barrier is to prevent moisture diffusion, and the job of an air barrier is to stop air leakage through differences in air pressure. A wall system should have one moisture barrier, but can have many air barriers. A moisture barrier can act as a very effective air barrier, but an air barrier does not (and should not) always stop moisture from diffusing.
- A wool sweater for example, is a good choice of natural insulation and will keep you warm when there is no air movement, but will allow the wind to howl right through it. A wool sweater with a raincoat will keep you warm but hold moisture inside and soak your insulation. A wool sweater with a windbreaker will keep you warm, stop the wind from stealing your heat, yet allow moisture to diffuse through it.
- So think of a windbreaker as an air barrier, and a raincoat as a moisture barrier. That is about as far as I can stretch the human to house analogy, hope it helps.
- Since warm air expands, there is more space between its molecules compared to cold air. Water moisture is found in that space. When warm air cools as it passes through your walls, it contracts and squeezes out the moisture, leaving you with condensation.
- In order to prevent condensation from forming, a moisture barrier should be placed on the warm side of your insulation to stop warm, moist air from condensing on a cold surface inside your wall.
- In cold climates like Canada, for most of the year the moisture barrier should be on the inside of the insulation. In hot climates like the southern U.S. for example, moisture barrier should be installed on the outside of the insulation, especially where there's air-conditioning involved to prevent condensation and mold.
- In both cases, the moisture barrier is tasked with preventing warm, humid air from shedding its moisture as it meets a cool surface, no matter which direction it is travelling.
- The most important thing to realize is that there is no fixed rule regarding moisture barriers. Building practices should always be determined by the climate in which you are building.

5.4 GROUND IMPROVEMENT TECHNIQUES

- The ground can be improved by adapting certain ground improvement techniques. Vibro-compaction increases the density of the soil by using powerful depth vibrators. Vacuum consolidation is used for improving soft soils by using a vacuum pump.
- Preloading method is used to remove pore water over time. Heating is used to form a crystalline or glass product by electric current. Ground freezing converts pore water to ice to increase their combined strength and make them impervious. Vibro-replacement stone columns improve the bearing capacity of soil whereas Vibro displacement method displaces the soil. Electro osmosis makes water flow through fine grained soils.
- Electro kinetic stabilization is the application of electro osmosis. Reinforced soil steel is used for retaining structures, sloping walls, dams etc. seismic loading is suited for construction in seismically active regions. Mechanically stabilized earth structures create a reinforced soil mass.
- The geo methods like Geosynthetics, Geogrid etc. are discussed. Soil nailing increases the shear strength of the in-situ soil and restrains its displacement. Micro pile gives the structural support and used for repair/replacement of existing foundations.
- Grouting is injection of pumpable materials to increase its rigidity. The jet grouting is quite advanced in speed as well as techniques when compared with the general grouting.
- Rapid urban and industrial growth demands more land for further development. In order to meet this demand land reclamation and utilization of unsuitable and environmentally affected lands have been taken up. These, hitherto useless lands for construction have been converted to be useful ones by adopting one or more ground improvement techniques.

5.4.1 Advanced Piling Techniques

- Ground improvement techniques such as stabilization, vibro flotation, dynamic compaction, stone column, compaction pile and compaction grouting are given the maximum importance in present days to adapt weak soil into the proper stable ground for different civil engineering projects.

5.4.1.1 Stone Column

- Stone column ground improvement involves adding vertical columns of stone into the ground to a depth of at least 4m below the ground surface. A layer of compacted gravel can then be put over the top of the columns, ready for the construction of new house foundations. The stone column method is quick to construct and can be done at any time of the year.
- It is a ground improvement technique used to improve the load bearing capacity and reduce the settlement of the soil. It is also called as granular columns or granular piles. This technique is also known as vibro replacement.
- In this technique dense aggregate column (stone columns) is constructed by means of a crane-suspended downhole vibrator.
- Stone columns are constructed in soil to increase the load-bearing capacities of the native soil. The upper portion of the compacted granular column is provided with a rigid central core such that vertical loads imposed on the composite column are transferred to a deeper level on the compacted column where the column operates more efficiently.
- A probe is centrally penetrated downwardly into the compacted granular column, and the resulting cavity is filled with cementitious grout to form a solid core after hardening. The grout may be injected into the cavity so formed in the compacted column at the bottom of the probe as the probe is being withdrawn in predetermined quantities metered in synchronization with the rate of withdrawal of the probe from the core cavity.

Function

- Installation of stone column improves ground by reducing soil settlement. Due to its higher modulus of elasticity than that of soil, it absorbs more load than soil and reduces overall settlement.
- Since applied load distributes in between soil and stone column in the ratio of their stiffness, the load carrying capacity of soil also increases.
- Stone aggregates are used to fill stone column. Water can easily pass into the stone column. So, stone column helps in excess pore water pressure mitigation and accelerates the consolidation process.

Displacement method

- In this method, boring is done by displacing nearby soil. The soil is displaced laterally, due to which engineering property of soil gets change.
- Hole can be made by driving casing or tube into the ground. Best example of displacement method is vibro float method which is being used widely.
- It is done by two methods.
- Wet, Top feed method.
- Dry, Bottom feed method.

Non-displacement method

- In this method, boring is done without displacing nearby soil. Although there will be some displacement that can be neglected. Example – bored rammed method.

Advantages

- Reduce total and differential settlement.
- Increase the bearing capacity of strata to make it possible to use shallow foundation hence, saving money and time.
- Accelerates the rate of consolidation in cohesive soil by providing drainage to water.

5.4.1.2 Vibro Floatation

- Vibro flotation is a ground improvement technique used at a considerable depth that by using a powered electrically or hydraulically probe, it strengthens the soil.
- The vibro flotation will compact the soil making it suitable to support design loads. It involves the introduction of granular soil to form interlocking columns with surrounding soil.
- The technique is used to improve bearing capacity and reduce the possibility of differential settlements that might be allowed for the proposed loads.
- Sometimes it is also referred to as Vibro compaction and the ultimate concept is to repack soil particles by joining them together improving soil's bearing capacity. The compaction of soil can be obtained in soils as deep as 200 feet. The risk of liquefaction in an earthquake prone area is also drastically reduced.

Vibro flotataion Techniques

Vibro flotation can be obtained by using three different techniques:

- **Vibro Compaction method -** This method allows granular soils to be compacted. This method is only used to compact sandy soils.
- **Vibro Replacement method -** The technique is used to replace poor or inadequate soil material by flushing out the soil with air or water and replacing it with granular soil. This can be used in various soil types such as clay and sandy soils.
- **Vibro Displacement method -** This procedure is used with no or small amounts of water used during the technique. The probe is inserted into the soil and it will displace it laterally as the new soil column is being formed and compacted.

Vibro flotation Advantages

Vibro flotation is one affordable way to improve ground conditions when a deep layer of inadequate soil is found. The technique is so simple that will not require the delivery of additional materials or additional equipment other than the probe and the equipment that has it installed. The vibro flotation process can offer the following benefits:

- When the process is done properly, it will reduce the possibility of differential settlements that will improve the foundation condition of the proposed structure.
- It is the fastest and easiest way to improve soil when bottom layers of soil will not provide good load bearing capacity.
- It is a great technology to improve harbor bottoms
- On a cost-related standpoint, it helps improve thousands of cubic meters per day. It is faster than piling.
- It can be done around existing structures without damaging them.
- It does not harm the environment.
- It improves the soil strata using its own characteristic.
- No excavations are needed, reducing the hazards, contamination of soils and hauling material out from the site.
- No need to manage table water issues, neither the permits required to manage water discharge and dewatering issues.
- The technique of vibro flotation can be adapted to each scenario and site.
- When vibro flotation is performed at a site, it will reduce the possibility of liquefaction during an earthquake.

5.4.1.3 Micro Piles

- Micropiles are small diameter piles (up to 300 mm), with the capability of sustaining high loads (compressive loads of over 5000 KN).The drilling equipment and methods allows micropiles to be drilled through virtually every ground conditions, natural and artificial, with minimal vibration, disturbances and noise, at any angle below horizontal. The equipment can be further adapted to operate in locations with low headroom and severely restricted access.
- The present invention relates to a method of joining a micro pile with an existing structure, and more particularly, to add a basic reinforcement according to occurrence of additional load due to settlement, flotation or extension of an existing structure, which is used for ground reinforcement in civil engineering and construction. The present invention relates to a method of combining micro piles with existing structures in the construction of micro piles that can be constructed in general.
- The construction method of the reinforcement micro pile according to the present invention includes:
  - (a) drilling a hole for inserting a micro pile into the foundation concrete;
  - (b) drilling holes for inserting micro piles into the foundation ground;
  - (c) inserting the micro pile into the holes drilled in (a) and (b);
  - (d) grouting or filling cement milk or mortar up to the foundation ground;
  - (e) expanding a portion of the perforated hole in the base concrete in the circumferential direction;
  - (f) fastening the load transfer member and filling the hole of the foundation concrete with cement milk or mortar.
- The size of the pile is determined by the load-bearing capacity of the ground and the size of rig that is able to access the piling location.
- Micropiles can also be used in combination with other ground modification techniques where complex site conditions and design specifications are present.

- In a typical installation process a high-strength steel casing is drilled down to the design depth. A reinforcing bar is inserted and high-strength cement grout pumped into the casing. The casing may extend along the full length of the pile, or it may only extend along part of the length of the pile, with the reinforcing bar extending along the full length.
- Drilling may be achieved by a removable bit, or by a sacrificial head to the steel casing.
- In some cases, the steel casing may be removed, or partially removed, and further grout pumped in at pressure.
- They can be installed in restricted access sites where there is low headroom by the use of lagging. This is where wood, steel or precast concrete panels are inserted behind the pile as the excavation proceeds so as to resist the load of the retained soil and transfer it to the pile. Greater capacity can be achieved by post-grouting within the bond length to increase frictional forces with surrounding soils.

There are advantages of micro piles include:

- They are small and relatively light.
- They are relatively inexpensive.
- They can be installed through almost any ground condition, making them suitable for installation in environmentally-challenging conditions, such as for wind turbine towers.
- The limited vibration and noise causes little disturbance.
- Piling rigs can be low-emission or even electrically driven.
- They can be installed while avoiding existing utilities, meaning that expensive utility re-routing is not required.
- They can be installed close to existing walls with limited headroom and congested site conditions.
- They have a high load capacity and an ability to resist compressive, tensile and lateral loads.

Applications of Micropiles for Ground Improvement

- For Structural Support and stability.
- Foundation for new structures.
- Repair / Replacement of existing foundations.
- Arresting / Prevention of movement.
- Embankment, slope and landslide stabilization.
- Soil strengthening and protection.

5.4.2 Soil Nailing

- Soil nailing is a ground stabilization technique that can be used on either natural or excavated slopes. It involves drilling holes for steel bars to be inserted into a slope face which are then grouted in place. Mesh is attached to the bar ends to hold the slope face in position.
- They are commonly used as a remedial measure to stabilize embankments, levees, and so on. Other applications for soil nailing include:
  - Temporary excavation shoring.
  - Tunnel portals.
  - Roadway cuts.
  - Under bridge abutments.
  - Repair and reconstruction of existing retaining structures.
- The main considerations for deciding whether soil nailing will be appropriate include; the ground conditions, the suitability of other systems, such as ground anchors, geosynthetic materials, and so on and cost.

- Although soil nails are versatile and can be used for a variety of soil types and conditions, it is preferable that the soil should be capable of standing – without supports – to a height of 1-2 m for no less than 2 days when cut vertical or near-vertical.
- Soils which are particularly suited to soil nailing include clays, clayey silts, silty clays, sandy clays, glacial soils, sandy silts, sand, gravels. Soil nailing can be used on weathered rock as long as the weathering is even (i.e. without any weakness planes) throughout the rock.
- Soils which are not well-suited to soil nailing include those with a high groundwater table, cohesion-less soils, soft fine-grained soils, highly-corrosive soils, loess, loose granular soils, and ground exposed to repeated freeze-thaw action.

Design considerations that will inform the design include:

- Strength limit: The limit state at which potential failure or collapse occurs.
- Service limit: The limit state at which loss of service function occurs resulting from excessive wall deformation.
- Height and length.
- Vertical and horizontal spacing of the soil nails.
- Inclination of the soil nails.
- Ground properties.
- Nail length, diameter and maximum force.
- Drainage, frost penetration, external loads due to wind and hydrostatic forces.

Some of the advantages of using soil nailing include:

- They are good for confined spaces with restricted access.
- There is less environmental impact.
- They are relatively quick and easy to install.
- They use less materials and shoring.
- They are flexible enough to be used on new constructions, temporary structures or on remodelling processes.
- The height is not restricted.

Limitations of using soil nailing include:

- They are not suitable for areas with a high water table.
- In soils of low shear strength, very high soil nail density may be required.
- They are not suitable for permanent use in sensitive and expansive soils.
- Specialist contractors are required.
- Extensive 3D modelling may be required.

Applications of Soil Nailing Technique:

- Stabilization of railroad and highway cut slopes.
- Excavation retaining structures in urban areas for high-rise building and underground facilities.
- Tunnel portals in steep and unstable stratified slopes.
- Construction and retrofitting of bridge abutments with complex boundaries involving wall support under piled foundations.

5.4.3 Vertical Drains

- This technique involves the installation of artificial drainage paths or vertical drains into soft compressible soils.
- The prefabricated vertical drain (PVD) and Sand drain used by for specially designed to work effectively under high settlements.

5.4.3.1 Sand Drains

- Vertical sand drains are used as an effective technique for construction of structures, such as oil or gas storage tanks or airports, sea ports, harbor works or highway or railway embankments or dams, over highly compressible soils.
- The principle of sand drains is to provide additional vertical drainage faces in the form of cylindrical holes filled with clean sand.
- Structures constructed over highly compressible soils, soft marine clay deposits suffer very large consolidation settlements, which have the potential to cause serious damage or failure of such structures. Such soils, located in the coastal regions such as Mumbai, Calcutta, Visakhapatnam, and others, extend to large depths of up to 15 to 25 m or more.
- Pile foundations may not always be an effective solution for such conditions. The surface of such soil deposits is so soft that it is difficult to use heavy construction equipment for construction of structures over such soils. The progress of construction is often delayed or halted for several weeks, shooting-up the project costs if conventional foundation techniques are used in such soils.
- Compacted embankments are usually constructed over such soils to provide a hard surface for construction and also to minimize the differential settlements.
- An ideal solution for this type of problems is to accelerate the rate of consolidation such that the potential settlements are completed before the construction of the structure.

Sand Drain is based on principles of rapidly and centrally dewatering day system.

- Sand Drain is a process of radial consolidation which increases rate of drainage in the embankment by driving casing into the embankment and making vertical boreholes.
- The holes are backfilled with suitable grade ofsand.
  Construction Process.
- Driven casing is withdrawn after sand has been filled.
- A sand blanket is placed over the top of sand drains to connect all sand drains.
- They accelerate the drainage; a surcharge load is placed on the sand blanket.
- The surcharge is in form of dumped soil.

Construction of Sand Drains:

- The sand drains installed below an embankment. A sand layer of 50-60 cm thickness is initially laid over the soft soil that not only serves as a drainage layer for the soft soil at top but also provides a working platform for construction of both sand drains as well as embankment. The sand layer also serves as a graded filter between the soft soil and the embankment and prevents loss of embankment material in to the soft soil.
- The embankment is built up to some height above the sand layer to facilitate partial consolidation of the soft soil. This facilitates the soft soil to gain some shear strength to support and sustain the sand drains. However, full height of the embankment is not constructed in single stage because the soft soil will not have sufficient shear strength to support the embankment, in which case the embankment would sink into the soft soil. After the embankment is constructed up to some height, the sand drains are installed.
- A circular casing or hollow steel mandrel, of about 25-cm internal diameter and 6-m height, is driven vertically into the soft soil up to the required depth through the embankment and sand layer. The soil in the mandrel is removed and the hole is back filled with clean sand. The mandrel is then removed. The embankment is then constructed up to the full height in stages. After allowing sufficient time for consolidation to complete, the required structure, such as pavement, airport, or oil storage tank, is constructed.

Advantages of Sand Drains:

The following are the advantages of using sand drains:

1. Provision of sand drains allows drainage of pore water in radial direction in addition to the drainage in vertical direction.
2. Since the permeability of soil in horizontal direction is usually several times larger than that in vertical direction, the rate of consolidation becomes considerably faster compared to conventional soil system.
3. With faster consolidation, the soil gains shear strength rapidly, allowing faster pace of construction and thus reducing the project cost.
4. The long-term stability of the structure is also increased significantly as the potential settlements are completed mostly before or during the construction.
5. Sand drains avoid potential problems during construction in soft compressible soils.

5.4.3.2 Pre-Fabricated Vertical Drains

- Prefabricated Vertical Drains (PVD's) are composed of a plastic core encased by a geotextile for the purpose of expediting consolidation of slow draining soils.
- They are typically coupled with surcharging to expedite preconstruction soil consolidation.
- Surcharging means to pre-load soft soils by applying a temporary load to the ground that exerts stress of usually equivalent or greater magnitude than the anticipated design stresses.
- The surcharge will increase pore water pressures initially, but with time the water will drain away and the soil voids will compress.
- These prefabricated wick drains are used to shorten pore water travel distance, reducing the preloading time.
- The intent is to accelerate primary settlement.
- Pore water will flow laterally to the nearest drain, as opposed to vertical flow to an underlying or overlying drainage layer.
- The drain flow is a result from the pressures generated in the pore water.

Advantages

- Decrease overall time required for completion of primary consolidation due to preloading
- Decrease the amount of surcharge required to achieve the desired amount of precompression in the given time
- Increase the rate of strength gain due to consolidation of soft soils when stability is of concern
- **Comparison to sand drains:** Economic competitiveness, fewer disturbances to the soil mass compared to displacement sand drains, and the speed and simplicity of installation. Also feasible to be installed in a non-vertical orientation.
- They have also outlined some technical advantages of PV drains compared to sand drains. They separate these advantages by sand drain type. There is displacement drains, which displace the soil as the drain is installed, and non-displacement, which does not displace the soil during drain installation.

They found the following advantages of PVD's:

Displacement

- Considerably less disturbance of cohesive soils during installation due to: smaller physical displacement by mandrel and top, and typically static push rather than driving.
- Installation equipment usually lighter, more maneuverable on site.
- Do not require abundant source of water for jetting.

Non-Displacement

- Do not require control, processing and disposal of jetted spoil materials; fewer environmental control problems.
- Field control and inspection not as critical.
- Definite potential for cost economy.
- Eliminate cost of sand backfill of drains, quality control problems and related truck traffic.
- Control and inspection requirements are reduced due to simplification of installation procedures.

General

- There is greater assurance of a permanent, continuous vertical drainage path; no discontinuities due to installation problems.
- PVDs can withstand considerable lateral displacement or buckling under vertical or horizontal soil movement.
- Faster rate of installation possible.
- Where very rapid consolidation is required, it is practical to install PVDs at close spacing.
- PVDs can be installed underwater and in a non-vertical orientation more conveniently.

Disadvantages

- If the compression layer is overlain by dense fills or sands, very stiff clay or other obstructions, drain installation could require predrilling, jetting, and/or use of a vibratory hammer, or may not even be feasible.
- Under such conditions, general pre-excavation can be performed if appropriate.
- Where sensitive soils are present or where stability is of concern, disturbance of the soil due to drain installation may not be tolerable.
- In such cases, sand drains installed by non-displacement methods or an alternate soil improvement technique may be more practical.

5.4.4 Thermal Methods

5.4.4.1 Soil heating

- It has been observed that heating or cooling shows certain marked changes in the soil properties.
- Heating and cooling have been extensively used as soil improvement techniques. Whatever be the mode of thermal stabilization we opt for it has the following needs:
  - Thermal evaluation of heat flow.
  - Heating or refrigeration system has to be designed.
  - Strength and stress-strain -time properties of the soil have to undergo performance analysis.

Concept of Heat Stabilization of Soil

- Like seepage or consolidation analysis of soil, a heat flow analysis can also be carried out. The transfer of heat in soil occurs by conduction, convection (free, forced, by thawing) and radiation.
- The most predominant mechanism of transfer is through conduction, which takes place in three constituents of soil which are soil solids, water (which may be in the form of a liquid, ice or moisture) and pore air.
- The phenomenon of heat conduction is influenced by soil thermal properties which are its thermal conductivity, latent heat of fusion, the heat of moistureization of soil water and the heat capacity of the soil.
- The behaviour of heat flow in soil is mainly governed by the latent heat of fusion of water on freezing and heat of moistureization of water on heating above 1000C. The latent heat of fusion can be defined as the heat amount that must be added to the unit mass of a substance to change it from liquid to solid or solid to liquid without any change in the temperature.

- It is noted that higher the heat input per mass of the soil (which should be treated), the greater would be the effect. A small increase in temperature will cause a strength increase in fine-grained soils due to the reduction of electric repulsion between the particles, pore water flow due to variation in thermal gradient and due to the reduction in moisture content because of increasing emoistureation rate.
- Therefore, it is found that it is technically feasible to stabilize fine-grained soils by heating. The following mentioned statement gives the temperature and the corresponding possible change in soil properties.

Applications of heat Stabilization of Soil

It is used in:

- Landslide stabilization.
- Improving soil undergoing collapsing.
- Mat foundation construction.
- Forming vitrified piles in place.
- Reduction of lateral stresses acting ton retaining walls.
- The method opted may be.
- Combustion to electrical.
- Investigations of microwave drying.
- Soil fusion by laser beam.

5.4.4.2 Soil Freezing

- Ground freezing is a soil stabilization technique carried out by continuously refrigerating the soil.
- There are numerous ways and technique of soil stabilization for deep excavation or tunneling. Some of them are grouting, heat treatment method and many more. One of the common and popular method used in the recent times is ground freezing technique.
- Ground freezing is a process of converting pore water or pores into ice by continuously refrigerating the soil.
- The water bearing soil is very losing and doesn't have enough compressive strength and shear strength to withhold its own loads. To increase these strengths and make the water bearing strata temporarily impermeable ground freezing method is used.
- This is usually done to provide structural underpinning, temporary support and prevent ground water flow into the site area. When the water-bearing strata is frizzed, the water in it transforms into ice which becomes a seal against the water and strengthens the soil.
- In grouting, extraneous materials are used. But in this method no extra material is required and after the work is completed, the soil reverts into normal state as before.
- This method can be use in any type of soil, regardless of size, shape or depth of excavation, soil or rock formation regardless of structure, grain size or permeability. However, it is best suited for soft ground rather than rock conditions.
- It is applicable to a wide range of soils but it takes considerable time to establish a substantial ice wall and the freeze must be maintained by continued refrigeration as long as required.

Conditions Where Ground Freezing is Most Effective

- Ground where penetrability by drilling, jet grouting, clamshell excavation, or other vertical cut-off tools is limited.
- Filled ground and ground containing man-made obstructions.
- Virgin ground containing cobbles, boulders, or an irregular soil/rock interface.
- Ground that has been disturbed due to unstable conditions or water inflow.

Principles of Ground Freezing

- The major principle of this method is to convert the water into ice by external freezing methods to create a water seal and strengthen the soil. The effectiveness of freezing depends on the presence of water to create ice, cementing the particles and increasing the strength of the ground to the equivalent of soft or medium rock.
- If the soil has doesn't enough amount of water to fill all to pore when they freeze, then it may be necessary to provide extra water so that the pores are complete sealed. This method is very effective in the places where the ground is made up of silts. Other methods of grouting cannot be undertaken due to very fine pores.
- The strength achieved by the ground after installing this method depends on freeze temperature, moisture content and the nature of the soil. After the initial freezing has been completed and the frozen barrier is in place, the required refrigeration capacity is significantly reduced to maintain the frozen barrier.
- When the ground water is transformed into ice, the expansion takes place which is negligible, around 9% expansion is observed which does not impose any serious stress or strain on the soil.
- As this is an artificial method of refrigeration, a uniform freezing can be imposed on any type of soil which offers great security when compared to various grouting methods.
- As in all ground treatment techniques, adequate site investigation is necessary to allow the best system to be chosen and to design the appropriate array of freezing tubes and select plant of adequate power.
- Once the freezing process has begun, monitoring is required to ensure formation of the barrier wall and to verify when freezing is complete. During the drilling process, temperature-monitoring pipes are installed to measure the ground temperature.

Types of Ground Freezing Methods

- The types of ground freezing methods used for temporary support of a tunnel heading are discussed below:
  1. Indirect Method.
  2. Direct Method.

1. Indirect Method of Ground Freezing

- This type of freezing method is commonly used in every place for the stabilization of tunnel headings. In this method, a secondary coolant is circulated through tubes which are driven into the ground.
- The figure shows the schematic representation of indirect ground freezing method. Two coolants are used in this method. The first is the Ammonia and the second coolant is Calcium Chloride.
- The ammonia is compresses in the freezing unit and passes to cool down. The condenser cools down the ammonia from gasses form to liquid form. The water is cooled down in the cooling unit.
- This liquid ammonia is passed and the secondary coolant is cooled down by the liquid ammonia. The cooled brine solution is passed to the tubes driven parallel to the ground. This tubes freeze and the surrounding area of the ground is also freezes. This cycle is repeated till the required ground area freezes.
- The time required to freeze the ground will obviously depend on the capacity of the freezing plant in relation to the volume of ground to be frozen and on the spacing and size of freezing tubes and water content in the grounds.

2. Direct Method of Ground Freezing

This method is further divided into two types:

1. Direct, by circulation of the primary refrigerant fluid through the ground tubes

- In this type of direct ground freezing method, only Ammonia is used in the freezing process. The process is same as that of indirect method, but here only one coolant is used.
- The ammonia is compressed and passed into the tubes driven into the ground. This ammonia freezes the tubes which in turn freezes the surrounding ground. Time required for this process is same as that of indirect process, but the efficiency is higher when compared to indirect method.
- The choice will depend on plant availability, estimates of cost and perhaps personal preference.

2. Direct, by injection of a coolant into the ground, such as liquid nitrogen

- This method does not require ant refrigeration plant. Ammonia is brought to the site under moderate pressure and stored at site in insulated tanks. Tubes are driven into the ground with the provision of return pipes to exhaust to atmosphere.
- There is an advantage for emergency use, i.e. quick freezing without elaborate fixed plant and equipment. This may be double advantageous on sites remote from power supplies. In such conditions the nitrogen can be discharged directly through tubes driven into the ground, and allowed to escape to atmosphere. Precautions for adequate ventilation must be observed.
- The speed of ground freezing with N2 is much quicker than with other methods, days rather than weeks, but liquid nitrogen is costly. The method is particularly appropriate for a short period of freezing up to about 3 weeks.
- It may be used in conjunction with the other processes with the same array of freezing tubes and network of insulated distribution pipes, in which liquid nitrogen is first used to establish the freeze quickly and is followed by ordinary refrigeration to maintain the condition while work is executed.
- This can be of help when a natural flow of ground water makes initial freezing difficult.

Advantages of Ground Freezing Technique

- Temporary underpinning of adjacent structure and support during permanent underpinning.
- Shaft sinking through water-bearing ground.
- Shaft construction totally within non-cohesive saturated ground.
- Tunneling through a full face of granular soil.
- Tunneling through mixed ground.
- Soil stabilization.

Disadvantages of Ground Freezing Technique:

- Very expensive.
- Needs continuously monitoring.
- Volume expansion of water during freezing, leading to soil heave and thaw settlement.

Important Points

- Zero energy homes are homes that produce as much energy as they use over the course of a year. Net zero energy is achieved through proper design, energy efficiency measures, and on-site renewable energy generation.
- These homes have zero energy bills, zero carbon emissions, increased home value, and improved health and comfort.
- Building and designing affordable zero energy homes – the ultimate in energy efficiency – involves 12 integrated steps that utilize commonly available building materials and equipment along with easy-to-learn building strategies. By following these steps, you can build a new home that is cost comparable to a conventional home.

- Twelve steps of zero energy building are as below:
- Start with Smart Design, Use Energy Modeling, Super-Seal the Building Envelope. Super-Insulate the Building Envelope, Heat Water Wisely, Use Highly Insulated Windows and Doors, Use the Sun for Solar Tempering, create an Energy Efficient, Fresh Air Supply, select an Energy Efficient Heating and Cooling System, Install Energy Efficient Lighting, Select Energy Efficient Appliances and Electronics and Use the Sun for Renewable Energy
- Success in building an affordable zero energy home always begins with the design.
- Depending on state rebates and incentives, zero energy homes may cost 5% to 10% more to build than a similar home built to code. However, those costs will be paid back within 5 to 15 years depending on each specific situation. Zero energy homes also can cost less to own beginning on day one and can have enhanced market value.
- There are a wide variety of existing zero energy homes built in all regions of the country, both in the far north and in the south. They range from single family custom homes, to development homes, apartment buildings and Habitat for Humanity homes.
- There are five steps to getting started on the path to creating a zero or near zero energy remodel of your home.
- Five steps are as below:
- Select a certified energy assessor consultant, select a contractor with experience in zero and near zero energy remodeling, select an expert in energy modeling for homes undergoing major structural remodelling, select a designer or architect experienced in zero or near zero energy remodeling and Save Money by Reducing Your Cost of Ownership.
- Homes that produce more energy than they use are called positive energy homes. The energy produced in excess of what is used in the home can be used to power an electric lawn mower or even an electric car.
- If there are financial restraints, consider building or remodeling a net zero ready home. If there is a lack of sunlight, consider building or remodeling a near zero home that has all the energy saving features of a zero energy home, but without the solar collectors.
- Green building is the practice of creating structures and using processes that are environmentally responsible and resource-efficient throughout a building's life-cycle from siting to design, construction, operation, maintenance, renovation and deconstruction.

Eco-friendly Construction: 8 Advantages of Green Building

- Low Maintenance and Operation Cost. Green buildings incorporate unique construction features that ensure efficient use of resources such water and energy.
- Energy Efficiency.
- Enhances Indoor Environment Quality.
- Water Efficiency.
- Better Health.
- Material Efficiency.
- Better Environment.
- Reduces Strain on Local Resources.
- Green Buildings provide financial benefits that conventional buildings do not. These benefits include energy and water savings, reduced waste, improved indoor environmental quality, greater employee comfort/productivity, reduced employee health costs and lower operations and maintenance costs..
- There nine "elements" are literally your green building blocks.

- Site planning and design. Affordable housing works best when residents have easy access to key services and transit.

  Community.

  Indoor Air Quality.

  Energy.

  Materials.

  Waste.

  Water.

  Commissioning.
- A green building rating system is a tool that evaluates the performance of a building and its impact on the environment. It comprises a predefined set of criteria relating to the design, construction, and operations of green buildings.
- Some of the more common green construction practices include: Using sustainable building materials like recycled glass and steel, as well as renewable materials like bamboo and rubber. Installing energy-efficient windows and doors. Using lower-VOC (volatile organic compounds) paints and stains.
- Green buildings preserve precious natural resources and improve our quality of life. There are a number of features which can make a building 'green'. These include: Efficient use of energy, water and other resources.
- The objective of green building concept is to develop buildings which use the natural resources to the minimal at the time of construction as well as operation. Green buildings emphasize on the resource usage efficiency and also press upon the three R's – Reduce, Reuse and Recycle.
- A high-performance building is a building whose energy efficiency and environmental performance is substantially better than standard practice. Although green buildings, on average, use less energy than conventional buildings, energy efficiency remains elusive.
- Building materials typically considered to be 'green' include lumber from forests that have been certified to a third-party forest standard, rapidly renewable plant materials like bamboo and straw, dimension stone, recycled stone, recycled metal.
- Indian Green Building Council (IGBC) Green Factory Buildings is the first rating programme developed in India, exclusively for the industrial sector. It is based on accepted energy and environmental principles and strikes a balance between known established practices and emerging concepts.
- The LEED rating systems are grouped into five main categories: Building Design and Construction. Interior Design and Construction. Operations and Maintenance.
- Green Architecture is largely driven by four goals.
- Reducing impacts on the Earth from constructing buildings and their materials (mining and harvesting, embodied energy, pollutants in manufacturing).
- Reducing impacts which arise during occupancy (fuel use, land pattern disruption, maintenance impacts).
- Eco-friendly, or ecological, construction is building a structure that is beneficial or non-harmful to the environment, and resource efficient. Eco-friendly construction has developed in response to the knowledge that buildings have an often negative impact upon our environment and our natural resources.
- Earth plaster and milk paints are excellent eco-friendly choices for building materials. Earth plaster is made out of mud and can be used as an interior or exterior finish. ... There are some great choices for eco-friendly flooring; cork flooring, bamboo, and recycled wood flooring are all excellent options.

Advantages of green technology to the environment are enormous, but the most important benefits are:

- ➢ It uses renewable resources that never depletes in nature.
- ➢ Environmental technology uses latest energy generation techniques.
- ➢ Offers environmental solutions for disposal and recycling of waste.

- Some examples of renewable energy sources are solar energy, wind energy, hydropower, geothermal energy, and biomass energy. These types of energy sources are different from fossil fuels, such as coal, oil, and natural gas.
- Mass Housing is an independent public authority that provides financing for the construction and preservation of affordable rental housing, and for affordable first and second mortgages for homebuyers and homeowners.
- With Mass Housing's new refinancing product, home owners can refinance up to 100% of the appraised value of their home, pay-off existing approved second mortgages, have the option of paying no points, and roll-in closing costs.
- With Mass Housing's new refinancing product, home owners can refinance up to 100% of the appraised value of their home, pay-off existing approved second mortgages, have the option of paying no points, and roll-in closing costs.
- Modular construction companies can build multi-family apartment buildings in a fraction of the time it takes using traditional stick-built techniques.
- Prefabricated construction is the practice of assembling a variety of components of a structure at a manufacturing site and transporting those sub-assemblies to the location of the construction jobsite. Prefabricated construction is sometimes thought of as a low-end and mass produced mode of construction.
- Costs of prefabs are typically cheaper than stick-built homes by an average of 10% to 20%. Faster Than Traditional Construction. Generally, it takes one to four months longer to build a traditional home than to construct a prefab. Flexible Size and Aesthetic.
- A modular home is not a mobile home; it is simply a home that is built off-site, as opposed to on-site. These homes are often called factory-built, system-built or prefab (short for prefabricated) homes. Modular and manufactured homes are not the same. Manufactured homes are not placed on permanent foundations.
- Precast is a construction product (mostly concrete) produced by casting concrete in a reusable mold or "form" which is then cured in a controlled environment, transported to the construction site and lifted into place. In contrast, standard concrete is poured into site-specific forms and cured on site.
- The study revealed that, on average, the precast concrete slabs were 23.22% cheaper than cast-in-place concrete construction and precast columns were 21.4% cheaper than cast-in-place concrete construction, especially for large scale production.
- Precast concrete is known for its high-quality and durability. One of the precast concrete advantages lies in its fireproof attributes. Precast concrete will not catch fire, and it will prevent the spreading of fire from room to room. It is a fantastic material to have a home or building made from for this very reason.
- Pre-Engineered Metal Building Construction is a cost-efficient alternative to conventional building design in commercial applications is the pre-engineered metal building system. Metal buildings consist of a structural steel framing system, supporting a metal roofing system and the wall panels of varying materials.
- Pre-Engineered Metal Building Construction. A cost-efficient alternative to conventional building design in commercial applications is the pre-engineered metal building system. Metal buildings consist of a structural steel framing system, supporting a metal roofing system and the wall panels of varying materials.

- Pre-engineered buildings are the new contemporary structures ideal for commercial purposes. PEB structures can accommodate industrial units that are used for manufacturing of consumer goods and products. Pre-engineered steel buildings are constructed at low costs and can be used for industrial applications extensively.
- This new photovoltaic technology is Solar paint. The new paint is composed of synthetic molybdenum sulphide.
- The paint will be based on dye-sensitized solar cells. Instead of absorbing sunlight using silicon like conventional solar panels, they use dye molecules attached to particles of the titanium dioxide pigment used in paints. That gives an energy boost to electrons, which hop from the dye into a layer of electrolyte.
- Solar paint, also known as photovoltaic paint, is exactly what it sounds like! It's a paint that you can apply to any surface that will capture energy from the sun and transform it into electricity.
- Acronym of BIPV (Building Integrated Photovoltaics) refers to photovoltaic systems integrated within an object. In the case of facade or roof systems the photovoltaic system is added to the building after it was built. These low powered systems of up to some 10 kW are usually integrated into the south facade.
- Photovoltaic glass (PV glass) is a technology that enables the conversion of light into electricity. To do so, the glass incorporates transparent semiconductor-based photovoltaic cells, which are also known as solar cells. The cells are sandwiched between two sheets of glass.
- Solar panels are made of tempered glass, which is also referred to as safety glass or toughened glass. It is created by thermal or chemical means. There are specific characteristics that make tempered glass suitable for the manufacturing of solar PV panels.
- Earthquake engineering include the analysis and evaluation of earthquake risk and the analysis and design of different types of engineering structures.
- Strength and ductility that are key consideration in earthquake resistant design.
- A "smart" base isolation strategy is proposed and shown to effectively protect structures against extreme earthquakes without sacrificing performance during the more frequent, moderate seismic events. The proposed smart base isolation system is composed of conventional low-damping elastomeric bearings and "smart" controllable (semi active) dampers, such as magnetorheological fluid dampers. To demonstrate the advantages of this approach, the smart isolation system is compared to lead-rubber bearing isolation systems. The effectiveness of the isolation approaches is judged based on computed responses to several historical earthquakes scaled to various magnitudes.
- For soundproof and fireproof inner walls, of a type comprising two parallel reinforcing metal plates mechanically interconnected to form an interspace there between and a filler of insulating material, constituted essentially of rock-wool like material arranged in the interspace. At least one continuous diaphragm extends parallel to the reinforcing metal plates and divides the filler into parallel layers. The diaphragm comprises a plate of solid refractory material having projecting edges projecting beyond the filler. The reinforcing metal plates have folded edges constituting section member like portions thereof. The projecting edges are in engagement with the folded edges of the metal plates. The folded edges are seam folded to hold the metal plates together.
- Layer of heat insulating material over which there is an outer rain and weather-proof layer which may, for example, be sheet metal, asphalt, tar or the like the latter materials being possibly combined with impregnated building paper or felt, a common characteristic of all of which materials is that they are substantially proof against diffusion of water moisture through the roof to the outside atmosphere as well as being proof against leakage of water from the exterior through the roof.

- The present invention discloses a composite profile for thermal bridge solid-web steel doors and windows and a manufacturing method, relating to a steel door and window profile and a manufacturing method. The profile provided by the invention consists of a solid-web steel profile made from standard steel member and a non-metallic heat insulation strip. The manufacturing method is that: firstly one to seven kinds of standard steel members are manufactured; secondly two standard members are seam welded into eight kinds of solid-web steel profiles; then two kinds of the solid-web steel profiles and one nonmetallic heat insulation strip are assembled into six kinds of composite profiles for thermal bridge solid-web steel doors and windows; and finally the coating and the assembly are carried out. Due to the small cross section, light weight, high strength, steel material saving and convenient assembly of the steel member, the profile provided by the present invention has the advantages of good heat insulation performance and high quality. The present invention also has the advantages of simple method, low equipment investment, standardized manufacture of the steel member, high efficiency and convenient popularization and application. Thus the present invention can be widely applied in the composite sections for thermal bridge solid-web steel doors and windows for the flat-opening door and window, the sliding door and window, the fixed window, the top-hinged swinging window and the manufacture.
- Construction process for the composite pile foundation which consists of the expanded belled base and the pile shaft, the construction process comprising the steps of:
  (a) Aligning the pile tip at the pile location and place the casing at the top of the pile tip;
  (b) Lifting the heavy hammer to enter the casing, repeatedly lift the heavy hammer to reach the given height and back press the casing, let it fall freely to ram the pile tip and the soil until a hole forms at the pile location;
  (c) Lifting the heavy hammers higher than the feed inlet of the casing and fill solid material from the inlet into the casing;
  (d) Ramming the above-mentioned solid material and repeat the filling and ramming operations;
  (e) Controlling the penetration of the last three blows;
  (f) Filling graded aggregate or stiff consistency concrete and repeatedly ram it to form the expanded belled base;
  (g) Pull out the casing;
  (h) Aligning the pile shaft at the pile hole and repeatedly back press and blow slightly with the heavy hammer to drive the pile shaft into the pile bole and fix and connect it with the expanded belled base;
  (i) Controlling the penetration of the last three bouts and form the pile. The invention offers the advantages of high bearing capacity, small settlement, short pile shaft, low engineering cost, simple construction, short construction period, easy quality control etc.
- Compacted granular or stone columns are constructed in soil to increase the load-bearing capacities of the native soil. The upper portion of the compacted granular column is provided with a rigid central core such that vertical loads imposed on the composite column are transferred to a deeper level on the compacted column where the column operates more efficiently. A probe is centrally penetrated downwardly into the compacted granular column, and the resulting cavity is filled with cementitious grout to form a solid core after hardening. The grout may be injected into the cavity so formed in the compacted column at the bottom of the probe as the probe is being withdrawn in predetermined quantities metered in synchronization with the rate of withdrawal of the probe from the core cavity.
- The present invention relates to a plain concrete compaction system applied to a ground hardening method using vibration compaction and a concrete column. More particularly, the present invention relates to a plain concrete compaction system suitable for a ground hardening method using vibration compaction and a concrete column, including a drilling rig unit forming an excavation hole up to a prescribed ground depth; a casing inserted into the

excavation hole; a concrete injection unit generating a concrete improver inserted into the casing; and a towing unit moving the concrete injection unit and the casing upwards. According to the plain concrete compaction system suitable for a ground hardening method using vibration compaction and a concrete column, the drilling rig unit forms the excavation hole up to the prescribed depth. Once the casing is inserted into the excavation hole, the drilling rig unit is lifted to the ground and the concrete injection unit is inserted into the casing. The casing and the concrete injection unit are moved upwards by the towing unit and plain concrete is filled from the lower portion of the excavation hole.

- The present invention relates to a method of joining a micro pile with an existing structure, and more particularly, to add a basic reinforcement according to occurrence of additional load due to settlement, flotation or extension of an existing structure, which is used for ground reinforcement in civil engineering and construction. The present invention relates to a method of combining micro piles with existing structures in the construction of micro piles that can be constructed in general.
- The construction method of the reinforcement micro pile according to the present invention includes:
  (a) drilling a hole for inserting a micro pile into the foundation concrete;
  (b) drilling holes for inserting micro piles into the foundation ground;
  (c) inserting the micro pile into the holes drilled in (a) and (b);
  (d) grouting or filling cement milk or mortar up to the foundation ground;
  (e) expanding a portion of the perforated hole in the base concrete in the circumferential direction;
  (f) fastening the load transfer member and filling the hole of the foundation concrete with cement milk or mortar.
- Micro pile, foundation concrete, foundation ground, expansion part, load transfer member.
- The present invention, the first row thumb pile indentation step of entering the first row thumb pile 100 along the boundary line of the excavation region; A second row thumb pile indentation step of entering the second row thumb pile 200 at intervals outside the first row thumb pile 100; A connection step of connecting the first row of thumb piles 100 and the second row of thumb piles by the connecting member 300 so as to be integrally behaved; An excavation structure forming step of installing excavation region 1 to form an excavation structure by installing an excavation member 10 between the first row thumb piles 100; Soil nail installation step of installing the soil nail 400 for the soil structure; by presenting a two-row thumb pile earthquake method using a soil nailing, including the earth pressure of the rearing soil even when performing a deep trench work It is to be sufficiently resistant to, and to suppress the occurrence of the overall displacement of the structure.
- Freeze wells will be used to isolate the area for soil remediation. Freeze wells will form a frozen barrier around a treatment area. Frozen barrier is to prevent the fluid from flowing into the processing area. Further, freezing the barrier is to prevent the contaminants move to the external treatment zone. Frozen barrier will be used to enclose the whole of weeks' treatment zone. Further, freezing the barrier may be formed under or above the treatment zone. Freeze well is activated before soil remediation, the frozen barrier is formed when the start of the soil remediation. Soil may be performed by any type of soil remediation systems, including thermal soil remediation system. The heater of the thermal soil remediation system, without passing through the restoration when the barrier, will be disposed proximate ice barrier.

Multiple Choice Questions

1. Energy consumed by buildings is equal to energy produced by buildings are known as ......
   (a) energy saving buildings (b) zero energy buildings
   (c) non-renewable energy buildings (d) none of these
2. Which of the statement is not correct in case of zero energy buildings ?
   (a) Zero energy bills (b) Zero carbon emissions
   (c) Increase home value (d) No health and comfort in living
3. Cost of zero energy building is more initially than conventional building by ......
   (a) 50% (b) 100%
   (c) 10% (d) 70%
4. Green building is known as a building ......
   (a) green in colour (b) newly constructed building
   (c) energy efficient building (d) which is harmful for environment
5. Which material you will not advice for construction of green buildings during construction ?
   (a) Concrete (b) Bamboo
   (c) Recycled glass (d) Agro waste materials
6. I.G.B.C. means ......
   (a) Internal green building committee (b) Indian green building council
   (c) Italian green building congress (d) none of these
7. Earth plaster and milk paints are good choice as a eco friendly material and used for construction of ....
   (a) cement concrete road (b) green building
   (c) multistoried R.C.C. building (d) bridge construction
8. Which is not an example of renewable energy ?
   (a) Solar (b) Wind
   (c) Hydropower (d) Thermal
9. Prefabricated construction is suitable for ......
   (a) a bungalow (b) a building
   (c) mass housing (d) none of these
10. Which is not an advantage of prefab homes ?
    (a) Cost is more than conventional site construction
    (b) Speed of construction is faster
    (c) Flexibility
    (d) Aesthetically good
11. For construction of a commercial premises which of these is most suitable ......
    (a) pre engineering buildings (steel) (b) in situ buildings (R.C.C.)
    (c) masonry construction (d) eco friendly buildings
12. Solar paint has a a constituent ......
    (a) lead (b) molybdenum sulphide
    (c) copper (d) nickel
13. B.I.P.V. means ......
    (a) building integrated photo voltaics
    (b) building of international provision of voltaics
    (c) build internally using photo voltaics
    (d) none of these

14. Solar panels used to store solar energy are made-up of ......
    (a) wood (b) plastic
    (c) glass (d) copper
15. The main consideration of a earthqual proof building is ......
    (a) appearance must be good (b) cost should be low
    (c) high strength and ductile (d) good elevation
16. Rock wool is used in sound proof walls as a ......
    (a) construction material (b) aesthetic material
    (c) insulating material (d) strong material
17. Pile foundation is a type of ....
    (a) shallow foundation (b) deep foundation
    (c) spread foundation (d) steel foundation
18. Which of the following is not emphasiers in sustainable development ?
    (a) Electronic information (b) Economic prosperity
    (c) Community well being (d) Environmental quality
19. Sustainable development and green building construction is related to ......
    (a) use of recycled materials (b) use of reused material
    (c) use of renewable source of energy (d) all of the above
20. "LEED" means ......
    (a) long energy efficient development
    (b) leadership in energy and environmental design
    (c) leadership in environmental engineering design
    (d) none of these
21. A mechanism of minimization of green house gases emission is related to ......
    (a) carbon dioxide (b) carbon credit
    (c) carbon fiber (d) none of these
22. Which is not correct statement in relation to pile foundation ?
    (a) It is a deep foundation. (b) It is used to transfer heavy loads on soil.
    (c) It is used in soil of low bearing capacity. (d) It is a type of shallow foundation.
23. Which are 3 R's related to green building ?
    (a) Reduce, Reuse, Recycle (b) Reduce, Rotate, Recycle
    (c) Rotate, Redo, Recycle (d) None of these

Answers

1. (b)	2. (d)	3. (c)	4. (c)	5. (a)	6. (b)	7. (b)	8. (d)	9. (c)	10. (a)
11. (a)	12. (b)	13. (a)	14. (c)	15. (c)	16. (c)	17. (b)	18. (d)	19. (d)	20. (b)
21. (b)	22. (d)	23. (a)							

Appendix : List of Micro Projects

1. Prepare the chart showing software and their applications.
2. Prepare a report on advanced materials.
3. Prepare a model of any one construction material used in recent days.
4. Prepare a report on advanced machinery.
5. Prepare a model of any one construction machinery used in recent days.
6. Elaborate the process of mass housing, pre-engineered buildings etc.
7. Make posters showing site safety and its awareness.
8. Prepare the charts showing different types of safety rules and regulation of site.
9. Prepare a report on advanced construction techniques.
10. Prepare a report on waste management.
11. Collect the relevant information of recent techniques in ground improvement and prepare a report on it.
12. Prepare the charts showing different types of natural disaster, causes, effects and risk reduction measures for that disaster.

●●●